DNA and Family History

How Genetic Testing Can Advance Your Genealogical Research

Chris Pomery

THE DUNDURN GROUP

TORONTO

For Vince, Jill, Alexander and Edmund, who all had great fun stimulating me to start this book, and not forgetting a friendly cuff for the three O'Brien family cats – Eric, Ernie and the aged Roland – who have used every feline means to prevent me from accessing the keyboard to complete it.

Library and Archives Canada Cataloguing in Publication

Pomery, Chris
DNA and family history : how genetic testing can advance your
genealogical research / Chris Pomery.

Includes bibliographical references and index.
ISBN 1-55002-536-8

1. Genealogy. 2. DNA. 3. Genetics. I. Title.

CS21.P64 2004   929'.1'072   C2004-904460-5

1   2   3   4   5   08   07   06   05   04

Editorial, design and production by The Book Group, Somerset
Printed in the UK by Cromwell Press

www.dundurn.com

Dundurn Press
8 Market Street, Suite 200
Toronto, Ontario, Canada
M5E 1M6

Dundurn Press
2250 Military Road
Tonawanda NY
U.S.A. 14150

CONTENTS

List of Figures and Tables iv
Foreword by Professor Steve Jones v
Preface vii
Acknowledgements viii

Introduction 1

PART I NEW SCIENCE, NEW HISTORY

1 The Ancestral Message in Your DNA 8
2 The Unfolding Story of Human Migrations 20
3 A New Reading of Surnames 36

PART II HOW GENETICS HELPS GENEALOGISTS

4 Reading Your Ancestral Message 50
5 Y-Chromosome Study Scenarios 62
6 Single-Ancestor Studies 70
7 Surname Studies 79
8 Clan and Caste Studies 88

PART III THE DNA TEST ORGANIZER'S HANDBOOK

9 The DNA Test Checklist 98
10 How to Select a DNA Testing Company 104
11 How to Launch and Market Your DNA Study 116
12 How to Analyze Your DNA Results 122
13 How to Present and Publish Your Study 134
14 The Future of Genetic Genealogy 148

Accompanying Website 155
Glossary 157
Further Reading 164
Index 166

List of Figures and Tables

Figure 1 Simplified structure of a cell 9

Figure 2 The structure of DNA 10

Table 1 Climate change and human migration 24

Figure 3 Path of Y-chromosome DNA from Africa to Asia and Europe 26

Figure 4 Phylogenetic chart of the main Y-chromosome clades 33

Table 2 Sample 10-marker Y-chromosome result 50

Table 3 Public databases of Y-chromosome test results 53

Figure 5 The six steps in the DNA test process 98

Figure 6 Five key criteria when choosing a DNA test company 109

Figure 7 Y-chromosome tests/value for money 111

Figure 8 Additional criteria when choosing a DNA test company 113

Table 4 Comparison of test companies' key capabilities 114

Table 5 Common ancestry rule of thumb 132

Table 6 Sample table: participant data from the Pomeroy study 136

Table 7 Sample table: DNA results of Pomeroy study participants 139

Table 8 Sample haplotype frequency table from the Pomeroy DNA study 140

Table 9 Modified haplotype table showing YHRD distribution data 141

Figure 9 Simplified mutation tree chart 142

Figure 10 Phylogenetic chart showing Y-chromosome haplotypes and mutations 143

FOREWORD

Man, unique among animals, can live in the past. History is the cement that binds society together, and its continued popularity is proof that even in the days of the worldwide web many people still anchor their identity in the notion of shared descent. Nowadays most nations find their essence in politics rather than blood but even that rational idea rests on an insight into a common past.

What is true for nations is just as true for families. As a result, many enthusiasts spend years in the record books, and on the web, tracking down just who their own ancestors may have been and from whence they came.

Once, that job was hard indeed and the results were often ambiguous. Now, everything has changed. DNA is the icon of the twenty-first century – and the key to all those that preceded it. The double helix has turned each of us into a living fossil: we carry in its imprint the history of our forebears, subtly changed by mutation. From the genes we can trace the origin of our families, our nation and, in the end, of mankind itself.

The problem with family trees is that they have so many branches. Everyone has two parents, four grandparents, eight great-grandparents and so on, and even allowing for the fact that each of us is inbred (for we all share common ancestors to some degree) the numbers soon become impossible to deal with. I probably descend, through one branch or another, from William the Conqueror – but so do you, and so do millions of others, which slightly takes the gloss off the discovery.

One set of genes escapes that problem. They are on the Y-chromosome, inherited – like a surname – down the male line. The Y is an arrow of manhood that flies from 'Adam' to every man alive today. In principle all men (and their wives and daughters, for they too have fathers) can use it to search out one crucial ancestor out of millions.

Chris Pomery has written a book that tells them how. It is for everyone who is interested in their own past; and who is not? Genetics, like many sciences, is filled with jargon, which can be hard to translate. As a result, people sometimes find it hard to understand quite what the subject can and (more important) can not do. When it comes to the search for ancestors, *DNA and Family History* puts them right.

It concentrates on the tie between surnames and genes. My own surname, Jones, began only in the eighteenth century, when the English pattern of family names spread to Wales. Even worse, it had many origins, for it means simply 'son of John' (as, for that matter, do the names Evans, Jenkins and Johnson). However feeble the evidence of my second name, my genes have given me a family. My Y-chromosome is of a type common in Wales and Ireland but rarer on the eastern side of England, which proves that, in the male line at least, I spring from ancient Celts. That slightly cheers me up, but I am happy to accept that the descendants of Saxons or of Slovenes would be equally pleased to trace their own roots.

Genetics ties the human family together as much as it divides it. Its most remarkable contribution has been to show how much we are alike. Compared to our relatives – chimpanzees, gorillas, and so on – Homo sapiens is astonishingly uniform from place to place. The average genetic difference between, say, a tribe of Aborigines and the population of Wales is less than between two groups of chimpanzees living a couple of hundred miles apart in Africa. We have evolved in our minds far more than in our bodies, and our ability to understand the past is a central part of what makes us so much more than apes.

DNA opens a new window into history. It can tell anybody who wants to find out exactly where on the human family tree they belong. This book shows, in a very clear and straightforward way, how to start looking.

Steve Jones
Professor of Genetics
University College London

PREFACE

I first became interested in DNA testing back in April 2000 when I read a summary of the results of Professor Bryan Sykes' ground-breaking Sykes surname study on the BBC website. My unsolicited comments were graciously replied to and six months later the Pomeroy DNA surname study was well under way in Bryan's lab at Oxford University.

As I delved deeper I began approaching academics around the planet in order to understand more about the power of genetic analysis, regularly receiving emails offering both enlightenment and encouragement. While DNA studies are, like astronomy, genuinely an area where an amateur enthusiast can have a significant impact on humanity's overall understanding, I confess that the learning process has to date been largely one-sided.

Some of the material in this book grew out of a website, hosted first on Genuki and later on Rootsweb, which I set up in 2001 to track the growth of genetic genealogy. This site has been expanded, rewritten and given its own domain to coincide with the publication of this book. You'll find that each chapter has its own set of online resources that you can access through the website, including notes that expand on all of the topics indicated throughout the text. You can also keep track of this subject as it develops by subscribing to my email newsletter.

The website is also a feedback route for you to pass your comments to me. If you think I've erred or missed out something important – either in this book or in the resources online – then email me. I value hearing from readers but please let me beg forgiveness in advance if I do not guarantee to reply to every email individually.

- This book's companion website <www.DNAandFamilyHistory.com>
- DNA test price information <www.DNAandFamilyHistory.com/tests>
- Email me feedback: feedback@DNAandFamilyHistory.com

ACKNOWLEDGEMENTS

I specifically want to thank Professor Steve Jones of University College London for writing the Foreword to this book; Dr Mark Thomas, also at UCL, for a range of academic papers; Patrick Guinness for his insightful email correspondence, notes on the Irish clan study and his family's warm hospitality in Kildare; and Ann Turner, the indefatigable moderator of the Genealogy-DNA mailing list on Rootsweb.

I also acknowledge a significant debt to Stephen Oppenheimer of Oxford University for the insights in his recent challenging book on prehistoric migrations *Out of Eden*.

Many DNA study organizers have clarified my views during the past half-decade including Kevin Duerinck and Doug Mumma in the USA, and two members of the Guild of One-Name Studies, Orin Wells and Alan Savin. I'd also like to thank five testing companies for sharing information and insights with me. My personal thanks go to Howard Benbrook, Philip Dance, Patrick Guinness, Maurice Hemingway, Vincent O'Brien, Diane Wilding, my brother John and my niece Elizabeth, who commented on different versions as I was writing this book. Needless to say, the opinions in it, and any errors, are entirely my own.

I also thank the publishing team at The National Archives and at The Book Group, specifically Anna Hodson who copy-edited this book in record time and Sheila Knight, my patient commissioning editor, whose soft Lallans tones cloak a wit with a daggered edge, notably in relation to the word 'deadline', a concept I claim now to have finally embraced.

Lastly, my thanks to the more than a thousand Pomeroy family researchers across the globe who've collectively helped me, directly and indirectly, to build up the documentary data that underpins the Pomeroy DNA project, and to the late Tony Pomeroy who shared both his enthusiasm and his research when I first came calling.

INTRODUCTION

Men may have few uses for the humble Q-tip, but a few seconds spent using a small ball of cotton to pick up some skin cells on the inside of the cheek may prove to be the best investment that any man, curious to know more about where his ancestors and his surname came from, can make. The cheek cells that adhere to the cotton contain within them your personal DNA signature. Once inside a genetics lab the cells will be teased apart to reveal a simple coded message that has the power to reveal a great deal both about your ancestry and who you're related to. The power of DNA testing is about to change genealogy for ever.

The genealogist's primary focus is to trace direct family lines and to build up family trees. As our surname records our most familiar line of descent, this book will focus on those DNA tests that can confirm the transmission of DNA, in the same manner as a surname, from male ancestor to male descendant. From the outset it must be clearly under-stood that the DNA test that is used to track male descent and surname transmission – a Y-chromosome test – can only be taken by men. While several types of DNA test can be taken by women, they cannot take a Y-chromosome test simply because they don't have one.

For years, family historians and genealogists – I use these terms inter-changeably – have lived largely in the twilight realm of the archive office, quarrying for documentary records whose relevance to their personal quest may be years away from realization. Already in the past decade the internet has dramatically increased the amount of documentary data available for researchers, so much so that the newest generation now expect to turbocharge their genealogy by instantly accessing huge online transcription projects and databases of family trees.

Why then do we need DNA testing? Firstly, the further back in time we go from the present day the more incomplete the documentary evidence available to us becomes. Our DNA, however, changes only very slowly. My personal Y-chromosome signature is almost identical to that of all of my direct male ancestors and all of their male descendants throughout my extended family tree. Secondly, many sets of documentary data that are coming online contain inaccuracies (particularly in genealogist-sub-mitted data such as family trees). My DNA signature, however, does not lie. Thirdly, as the amount of documentary data that is available increas-es, I can use DNA test results to focus my research in ways that not only improve the accuracy of my findings but which save me time and money. And lastly, DNA tests offer the best chance I can see of accelerating me towards the genealogical researcher's nirvana: the point of closure where my research will broadly be complete.

DNA testing is not new. It has been used for two generations by academic palaeoanthropologists (who investigate how human populations have evolved both biologically and culturally) and palaeogeographers (who track the trans-continental migrations of early Homo sapiens). Many readers will have come across it through the forensic criminal investigation procedures much popularized on television. And just as the techniques developed for forensic analysis have been commercialized by companies offering reliable paternity tests, so the tests developed by the academic disciplines researching human prehistory have now entered the mainstream packaged for genealogists. So even if the practice of DNA testing by family historians dates only from the late 1990s, the science upon which genealogical DNA test kits are based is really well established.

At the outset it's worthwhile to reassure readers that the type of DNA test that is offered to genealogists can tell you nothing definitive about your medical history or your skin colour. There is no DNA test for racial identification as 'race' is a social construct. Even the tests that identify a particular British or African tribe that your ancestors belonged to do so only in approximate ways.

Many people are reflexively afraid of the whole concept of genetic testing. But while one has to handle DNA issues sensitively and remain alert to the fact that a test result may reveal a different genetic heritage to that which the participant expected, this should not cause you to hesitate.

The good news is that while the price of the basic genealogical DNA test has been dropping since it was first commercialized, the value for money represented by each grade of test on offer has increased even faster. Compared to the travel costs of a research trip to a county record office or to access genealogical databases online, a DNA test is easy to arrange and represents an excellent long-term research investment.

My personal view, based upon watching the growth of the sector during the past five years, is that DNA testing offers genealogists a radical new tool that will increasingly reinvigorate our family history research in a very deep way. In fact, I think that DNA testing – and the 'genetic history' it reveals – is set to become the third major foundation upon which family historians build their genealogy, complementing both oral history and documentary history. I see it provoking something like a revolution in the genealogical world so that in a couple of years' time a DNA test will be seen as a valuable and sensible first step for many people venturing into family history research for the first time. More optimistically, I also think the new set of data revealed in our DNA results has the power to transform genealogy within the growing business of popular history from a slightly dowdy poor cousin into a much loved and respected uncle.

The impact of DNA testing on genealogy

Mainstream historians increasingly realize that genealogy has a respectable – potentially crucial – role in expanding our understanding of the social history of the last millennium. Long-term community and family studies require their subjects to be accurately linked by their genealogy over many generations. Furthermore, the study of surnames, a huge and contentious field, has barely got beyond the first stages of basic etymological categorization. Only a very few complete surname genealogies have so far been reconstructed in a way that reveals the full complexity of the economic development and social fortune of all their members. For me this is one of the most exciting implications of DNA testing, that it will help usher in a 'coming of age' that uses genealogy to redefine the study of surnames through the reconstruction of many whole-surname family trees, with the prospect of stimulating a new kind of integrated social history.

DNA testing should turn out to have an even greater impact on genealogy than the development of the internet. While the web has massively increased our access to a great deal of documentary data, much that is proliferating online is outdated or of dubious accuracy. Exactly the opposite process is, however, at work in the field of genetic testing where, as the corpus of DNA results increases, so the value of each individual test result becomes more meaningful over time. Put simply, the more DNA data we have to compare, the more information we can read within each man's individual DNA result. Our collective knowledge of how to interpret individual test results is improving very fast, and that means that men taking a DNA test today can expect to be able to learn more and more from their result in the years to come even without doing any further research and without paying a cent more. The internet is an enabling device that has helped me do things quicker; DNA testing is transforming my research by allowing me to see all my data in an entirely new way.

How this book works

This book, and its accompanying website, is designed to help those of us (myself included) who do not have degrees in genetics or statistics to be able to appreciate the full value of a DNA test result. It will help you to decide what type of study you want to start and which DNA test is the right one to buy in order to try to solve your particular research problem, and it outlines what kind of insights you can expect to learn from your results.

The focus throughout this book is on Europe, specifically on families and surnames associated with the British Isles. This is because many researchers in the New World and the former British dominions find that their key purpose in taking a DNA test is to identify their historic family in the Old World. The principles described here are, however, true for any other location or type of genealogical project.

I've laid out the chapters in this book to make it easy for someone who knows nothing about either genetics or DNA testing to gain the maximum benefit by working through the book sequentially.

The book itself has three main goals, namely:

1. To explain what insights a DNA test can reveal to you as a genealogist.
2. To describe what other people have been able to achieve through DNA testing.
3. To advise you how to organize your own DNA test study.

The book is divided into three parts: **Part I** explains what a DNA test is and sets the context for you to understand your results by explaining how humans spread across the globe through a series of prehistoric migrations and how historians classify surnames. **Part II** explains the types of research scenarios that a DNA test can help you to resolve and gives examples of successful test programmes for each type of scenario. **Part III** contains the practical handbook-type information that explains how you can move from understanding your own personal DNA test result to become the organizer of a wider DNA study. These chapters explain how you can organize a DNA test project for your surname, how to choose the best testing company to achieve your research goals, and how to interpret, publish and present your DNA study's results. The final chapter looks at the future of genetic genealogy and hazards a few guesses about how the field might look in a few years' time.

Throughout the book I've tried to introduce the science behind DNA testing where it is relevant in the narrative rather than in a single hard-to-digest chapter at the beginning. That said, Chapter One should give you all the background science that you need to understand and enjoy this book. In a book this size much of the science – the physical analysis of the DNA sample, for example – is too complex to be explained in detail. Interesting though it is, it's not vital that you as a family historian should know it. This book is written on the assumption that you prefer to take your science shaken not stirred; the companion website to this book supplies a host of references so that you can follow these specialist areas up yourself as you wish.

At the back of the book I've created three sections that I'm sure you'll find helpful: an appendix describing the resources on this book's companion website which holds information updated since the publication of this book, background information and a set of checklists for DNA study organizers; a jargon-busting Glossary of scientific terms; and a list for Further Reading that is aimed at those readers who want to go on and explore the major topics in more detail.

Genetic testing is a rapidly developing business. In particular, test prices change regularly as new tests are introduced so be forewarned that the figures cited in this book may become out of date.

How the website works

The slim volume in your hands represents the first book-length attempt to summarize what has been learned about DNA testing, as it affects family historians, during the past five years. The field is changing so fast though that by the time the second edition comes out some facts and ideas will surely have been refined by further discovery. The companion website is designed to help you monitor this process as it happens so check it out at <www.DNAandFamilyHistory.com> for further details and comment. You can also use the website to buy the books described in the Further Reading section.

How to access the best test prices

To help you choose the right testing company for your project and to access the latest prices, please go online to <www. DNAandFamilyHistory.com>.

DNA testing is a daunting topic for non-scientists like me. My view is that retail-level DNA testing of the type described in this book will only become truly mainstream when the science and statistics that are an integral part of it are packaged in a simple way and made highly accessible. The good news is that some of the testing companies are already doing a very good job of explaining to their clients how their DNA results should be interpreted.

DNA testing is today still in its infancy, yet it is poised to enter our daily lives. I'm sure that in a few years' time we'll look back from a future where we carry a biometric, genetic datacard in our wallet as the proof of our personal identity to realize that the simple DNA test we took as a genealogist was our first direct encounter with this brave new world.

I certainly hope that by the time you finish this book you'll feel inspired to have a DNA test done.

More info at <www.DNAandFamilyHistory.com>

- DNA test price offers from the testing companies

- Up-to-date checklists for study organizers

- Lists of books, papers and online resources to help your DNA study

---

PART I

NEW SCIENCE, NEW HISTORY

1   *The Ancestral Message in Your DNA*

2   *The Unfolding Story of Human Migrations*

3   *A New Reading of Surnames*

CHAPTER 1

The Ancestral Message in Your DNA

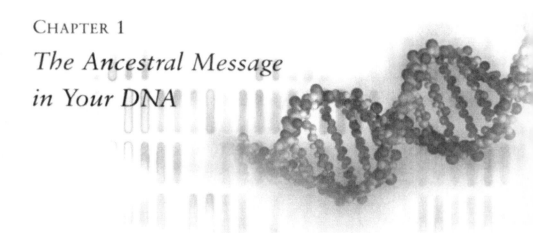

Reading DNA

My DNA is unique. Even so, if the entire genetic sequence contained within my cells could be laid out for inspection it would look similar to each of my brothers' DNA, to my parents' DNA, and also to yours. The crucial question is: just how similar?

A test from a forensics lab probing my DNA in just 13 places can differentiate me from everyone else on the planet with odds calculated at one in a trillion, a figure roughly 150 times larger than the number of humans presently on the planet. A simple paternity test available on the internet would check my DNA sample with that of a suspected child of mine to demonstrate to a high degree of probability that we are directly linked.

Another set of DNA tests can reveal a wealth of detail about the 'deep ancestry' of my prehistoric origins. And most recently still, a simple test has been devised for genealogists that will do two things to assist me as a family historian. Firstly, it quantifies the degree of relatedness that my DNA sample has with those of other men to reveal whether we share a common paternal ancestor within any recent time-frame.

Secondly, if the degree of relatedness between myself and those other men is significantly strong, it provides me with an estimate as to how many generations ago our common paternal ancestor probably lived. This should tell me whether our common ancestor could have existed in the time period since surnames came into use and thus whether a documented link could potentially be ascertained through further documentary research.

What do DNA tests measure?

At the simplest level, DNA tests look at certain places in selected human genes and ascribe a numerical value – called an allele value – to the pattern of DNA that they read there.

Our genes are transported inside chromosomes that are themselves found inside the nucleus of each of the 10^{14} (or 100 trillion) cells in the human body, all of which originally stemmed from a single cell. Human genes, which are simply a form of code designed to trigger the production of the proteins needed to create a replica cell, are made up of two types of code: sequences of active code that perform this useful function and redundant 'junk' code that appears to perform no useful genetic role at all. Perhaps surprisingly, in higher organisms such as humans more than 90% of gene sequences are made up of this 'junk code'.

Figure 1 Simplified structure of a cell

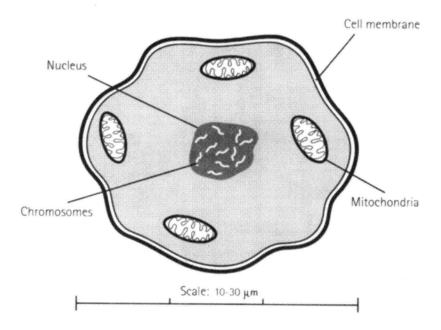

The nucleus of a cell contains 23 pairs of DNA-bearing chromosomes that collectively define genetic make-up. The body of the cell (the cytoplasm) also contains several mitochondria, and numerous other organelles.

The actual genetic information in our genes is held in large long-chained molecules known as DNA, specifically in subunits of DNA known as nucleotides. These are in turn made up of a number of 'bases' arranged in complementary pairs – known as base pairs – that are linked together in the classic double-helix pattern we now visually associate with DNA. The helix fits together because the four constituent bases – adenine, represented as A, cytosine as C, guanine as G and thymine as T – will only bond as a pair in single way, A always with T and C always with G.

Figure 2 The structure of DNA

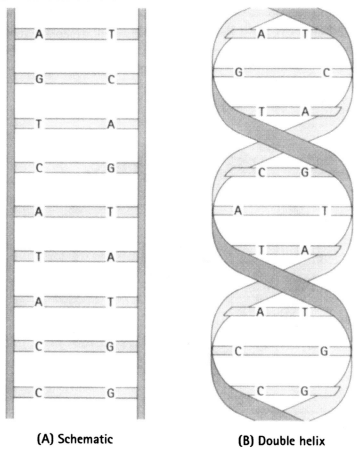

(A) Schematic   (B) Double helix

DNA's twin strands consist of linked nucleotides formed of base pairs of adenine/thymine and cytosine/guanine nucleobases. These two strands are wrapped round each other in the double-helix pattern.

The function of DNA is to summarize the instructions for cells to make proteins and to set up a method to make copies of the cell itself so that the new cells created by cell division will carry information that is identical to the original cell. During the cell replication process, which is extremely complex, some differences do arise when slight mutational changes in the DNA code – also known as genetic polymorphisms – are created. The mutational changes in the active coding regions account for all the differences that we can see between each human being, including genetic traits such as red hair and genetic defects such as inherited diseases. Mutations in the non-coding 'junk' DNA regions cause no known effect on us as humans. They do, however, create an indelible 'DNA signature' for each of us as individuals, a signature that contains within it a clear record of our ancestry and one which genealogical DNA tests are now designed to read. The mutational changes in the non-coding regions can take one of several forms. The most common that we need to know for our purposes are single nucleotide polymorphisms (SNPs, pronounced 'snips') and short tandem repeats (STRs).

Single nucleotide polymorphisms occur at only a single nucleotide at a specific position in a particular chromosome. Most importantly, this change occurs only once in a single individual. SNPs are therefore akin to an on/off switch that was thrown at a particular moment in time and that has not changed since. When you are DNA tested on a SNP marker you will in effect have either a positive or a negative result. The group of descendants that each SNP defines based upon those two results is known as a clade or haplogroup.

An STR polymorphism occurs when the length of a sequence of junk DNA code changes as the result of a mutation during the replication process. What happens is that a few extra base pairs of code are added to the sequence or are left off. STR mutations can happen at any time at the single position (or locus) that they are each associated with in the non-coding regions of our DNA, and they regularly do mutate in different individuals. As the actual number of base pairs repeated at each locus is highly variable, the result from a test of any particular STR will show one of a range of values. Comparing the values of a set of STRs together turns individual STRs into a very useful tool indeed to match different humans together.

STRs are extremely common throughout our genes but only a few are measured in the commercial DNA tests that genealogists will encounter.

A set of allele values measured at a number of different loci is collective-ly known as a haplotype while the loci of both SNPs and STRs are referred to colloquially as 'markers'. (Technically STRs are a type of vari-able number tandem repeat (VNTR) marker. There are a number of other technical terms that you might meet in other books and articles, but I've put information on these on the website so as not to weigh this narrative down more than is necessary.)

One word of warning: when you meet the term 'haplotype' you have to be careful to ask yourself about the degree of resolution that it describes. A haplotype can be defined as comprising as few as 4 or in excess of 40 markers. However, there's a great deal of difference about how much information you can read into a 4-marker haplotype com-pared to a 40-marker haplotype. We will look at this issue in Chapter 12 when we consider how to analyze DNA results and haplotypes.

Types of DNA test on offer

DNA testing has been widely used by academics over the past two decades. The field most closely related to genealogical genetics is an area known as phylogeography. By studying the distribution of haplogroups in present-day populations, the phylogeographer infers how those popu-lations may have migrated there from the original home of mankind in Africa. As we'll see in the next chapter, their studies have revealed the var-ious routes of the prehistoric migrations made by mankind across the planet during the past 200,000 years, information that creates an inter-esting context for our own ancestral investigations.

The DNA tests used by family historians to generate a haplotype result are designed to shed light on two key unknowns:

1. The geographical origins of our earliest known ancestors.
2. The genetic relatedness we share with other humans through the ancestors we share in common.

As already mentioned, the most useful DNA test for genealogists is one that tracks the male-to-male transmission of DNA from father to son, mimicking the manner in which surnames are handed down from one generation to the next. This male-only DNA test is a Y-chromosome test

– sometimes described as a Y-line or Y-test – as it measures STRs on the sex-determining Y-chromosome. A Y-chromosome test can only be taken by men for the simple reason that women don't have one.

There is a test that tracks the female line of descent, the mother-to-daughter transmission of genetic data. This test is described as a mito-chondrial DNA test, and is named after another part of the human cell that the test analyzes. While mothers pass on their mitochondrial DNA (mtDNA) to all of their children, whether male or female, the mtDNA signature is only passed on to the third generation down the female line. A man takes his mtDNA signature from his mother, not from his father.

Finally, there are also some unusual tests that target the chromosomes other than the Y, what is known as autosomal DNA.

This book will focus almost exclusively on the Y-chromosome test simply because this is of paramount interest to genealogists who are building surname-based family trees.

Y-chromosome tests

The main function of the Y-chromosome is very clear: it determines the sex of the future child. Geneticists like it because it has only two active genes on it compared to the average of 1,500 on each of the other 22 chromosomes. Most of the coding space within the Y-chromosome is made up of junk DNA, which means that it is a very good place for geneticists to look for the useful STR polymorphisms. Even better, as it can only be handed down from male to male, its transmission from generation to generation tracks the male line of descent and surname transmission.

DNA tests designed for genealogists look at specific STR markers on the Y-chromosome and measure the number of base pair or sequence repeats that are present in the code sequence at each of those loci. Typically the value at any given marker will increase or decrease by just one, or sometimes two, repeats every time a mutation takes place during the copying of the DNA between two generations. Mutations are, however, relatively rare. This ensures that when the values of a number of different STRs are compared it is relatively easy to say how related the two tested men really are.

If you pool together or aggregate the STR results of a large number of men, each STR marker will show a range of values that cluster around the modal – the most common – allele value for that marker. For example, for the marker known as DYS 19 the modal value among European populations is 14 while extreme values ranging as low as 10 and as high as 19 have been found in modern-day populations.

Just to make it complicated there are a few special markers that test companies use which do not produce a result that is a single allele value. These are paired markers, which produce a pair of values, and multi-copy markers, which produce multiple linked values. I've put detailed notes on these special markers on the website.

Mitochondrial DNA tests

Tests of mitochondrial DNA – generally written as mtDNA – have long been used in migration studies. However, they are of limited use to genealogists because mitochondrial DNA is passed from generation to generation only down the maternal line from mother to daughter. Phylogeographers like it because while mtDNA is very small – just 0.0003% the length of the Y-chromosome – it has relatively more polymorphisms and it mutates 10 times faster than the DNA found in a Y-chromosome. This makes it even more useful to pinpoint changes in, and movements of, human populations over time.

The international model for mapping mitochondrial DNA, to which all mtDNA results refer, is known as the Cambridge Reference Sequence.

Autosomal DNA tests

Mitochondria and the Y-chromosome have been the first choice genetic test areas for phylogeographers for one key reason: they both replicate their DNA from generation to generation with just the minimal changes caused by random mutation.

Autosomal DNA, on the other hand, which is the DNA in the 22 chromosomes other than the Y-chromosome, goes through a complex additional process during the cell replication process that geneticists call recombination. When a new human is created it shares DNA from both of its parents. During the process to create a new being, some of the parents' individual genetic inputs are shuffled about, the net effect being to make the two direct ancestor's genes a bit less distinct in the new child. Clearly all the children of the same two parents potentially share the same genetic input from their combined ancestors, though in reality the actual proportion they share from, for example, any of their four grandparents is in the range 0–50%. In practice, given that the genome is so large, the percentage from each grandparent will be around 25%. The problem is that, unlike the non-recombining Y-chromosome and mitochondria, it is extremely difficult (and, from a genealogist's point of view, impracticable) to identify what genetic input has come from which individual ancestor when analyzing autosomal DNA.

Trace your family tree back 12 generations and theoretically you could have a total of 4,096 unique ancestors providing input into your DNA. Expand that time-frame by a further 8 generations to 20 generations from the present day and the number of potential ancestors rises to over a million. These numbers suggest that not all of these ancestors can be unique contributors to your DNA signature which will include DNA from non-unique sources, from first-cousin marriages to very distant genetic relationships. The key point is that during the recombination process the DNA is jumbled up. The only routes by which the DNA inputs can be individually tracked back from descendant to ancestor are by using the Y-chromosome and mtDNA tests.

There are some autosomal DNA tests on the market aimed at family historians. The results they produce purport to indicate the percentage of your total DNA associated with particular ethnic modern populations. While it might be an interesting talking point that, for example, my DNA is 65% Indo-European, this information is of no use to me as a genealogist. In fact I think it is best to think of the mtDNA and autosomal DNA tests as 'ancestor heritage' or 'deep ancestry' tests so as not to confuse them with Y-chromosome genealogical DNA tests.

DNA testing and surname studies

The first study that used DNA tests to investigate the common ancestry of a group of people bearing the same surname reported its results in a scientific journal in the year 2000. The study, carried out by Oxford University's Professor Bryan Sykes, was similar to the tests carried out by the phylogeographers except that instead of measuring slow to change

SNP markers he tested a range of the faster-changing STR markers. What made the study groundbreaking was that he chose to test a predefined segment of the total British population: men with the surname Sykes.

Although several later migration studies in Britain have included a surname element, the application of DNA testing in the field of genealogy is still so new that as of mid-2004 no follow-up scientific paper focusing purely on surnames has been published. Despite the infancy of this type of application, a few labs offering DNA tests designed specifically for genealogists have already sprung up in Europe and the USA. From a standing start, the worldwide market for genealogical DNA tests has already grown to well over US$1 million annually and has given every indication that it will continue to grow as fast or faster in the next few years.

A first glance at DNA results

The result that a testing company will supply as the output of a Y-chromosome DNA test will be a string of numbers, one figure per marker or locus, collectively called a haplotype or DNA signature. The DNA result for any single STR marker does not tell you a great deal. If there are, say, nine possible numerical allele values for a particular marker, perhaps two-thirds of all people tested might have one or other of the two most commonly found results. Just a few people will have one of the rarer results. What is more significant is the pattern of a set of results derived from a range of markers. Put another way, the individual markers' results are a code but their combination together is the message.

Genealogical DNA tests work because when a large number of STRs in each individual are sampled you can more accurately describe the statistical probability that two people's results are close enough to being identical that they share a single common paternal ancestor, i.e. that they are both directly related through their fathers up through many generations along the direct male line back to the male ancestor they share in common.

Any two haplotypes can be compared with each other at any resolution of markers. You can compare as few as four markers – as done in the original Sykes study – or enhanced at a much higher resolution with 10 times as many markers. Indeed, the commercial tests currently offered to family historians use between 9 and 43 markers. At almost any marker resolution, an exact match where the two men's haplotypes are identical is significant: it suggests they share a common male ancestor. Beyond that simple statement there is a great deal of statistical analysis that can be performed on the test results, but from the point of view of most family historians very little of it is of practical use in assisting us in our primary goal: the reconstruction of family trees. Anyone who wants to tackle the statistics will open up a fascinating field to explore. The rest of us meanwhile can happily move forward using some basic rules of thumb. What you need to bear in mind is that the analysis of potential linkages between the DNA test results of different participants is always expressed in terms of the degree of probability that those linkages are true. While this may sound infuriatingly imprecise it does in fact make the results easier to use.

How are family historians using DNA test results?

The genealogist's first task is to use the results from men who've taken a Y-chromosome test to cluster together these participants into 'genetic families'. To give a simple example, a good outcome for, say, three men taking a Y-chromosome test is that the first and the second learn that they are highly likely to be directly related through a single common ancestor but that neither of them is related to the third man in any time-frame that makes sense to a genealogist. Clearly the owners of two identical DNA signatures have a great deal in common with each other just as two men with completely dissimilar DNA signatures do not. The most interesting analysis occurs in the grey area in between, in the area where the two men might be related within a genealogical time-frame but equally possibly might not. Any reader that I've not already made sleepy will be quick to spot the advantages that can be gained by pooling the results of several dozen men bearing the same surname. Lines of descent existing in family trees can be confirmed and people who do not know their family history can be associated with an already documented tree. The flip side is that some documented links may not be confirmed by the DNA results and some lovingly created family trees may need to be re-examined and even rewritten.

Whatever answer the analysis suggests, and even though we may have to couch our understanding of some results with a number of qualifications, DNA test results do advance our genealogical knowledge and help us prioritize further research.

How the study of prehistoric human migrations helps genealogists

One of the exciting features of the type of DNA test that family historians take is that every individual's result adds to our collective knowledge and makes every other individual's test result a little more meaningful. The Y-chromosome results that we receive back from a DNA test will surely grow in value in the years ahead as progressively more and more Y-chromosome data are collected from an expanding number of people around the world.

All of our new data are building upon data that have already been pieced together in scientists' attempts to understand both the origins of humanity and the way in which that ancestral message is encoded in our DNA. The next chapter looks at the how academic geneticists around the world are trying to chart the prehistoric migrations that led to the spread of mankind around the world. These prehistoric migrations are precursors to the more familiar migrations of the modern historical era, and indeed it would be hard to understand how, for example, the Vikings came to spread across Europe unless the context of the post-Ice Age population expansion back across the freshly uncovered continent more than 10 millennia earlier is first explained.

The DNA tests used by genealogists today are essentially the same as those used in many of these long-established academic programmes, and we are exceptionally fortunate that much of the background data collected are available in one form or another – often for free on the internet – in formats that non-academics can use at first hand to inform our own genealogical studies.

The next chapter will look at our 'deep ancestry' and try to answer the following questions: how did humans get to where we are today and what does this history of trans-continental migration mean for genealogists?

More info at <www.DNAandFamilyHistory.com>

- Genetics for beginners

- Y-chromosome tests, MtDNA tests and autosomal DNA tests

- Haplotype resolutions

Chapter summary

- The Y-chromosome DNA test is designed to help you to work out which men, with whom you share a surname, you are likely to be directly related to

- The Y-chromosome DNA test is offered by companies at low, medium and high resolutions; the high resolution test is the preferable, and often essential, choice

- As more people take DNA tests and their results are aggregated, our collective ability to interpret each individual test result improves

- As a side benefit, DNA tests can help you to identify your 'deep ancestry', the geographical origins of your direct paternal ancestor

CHAPTER 2

The Unfolding Story of Human Migrations

Footprints in the sand

Around 125,000 years ago, during a relatively warm climatic period that lasted 15,000 years, a group of human beings butchered a large animal on a beach near Abdur on the Eritrean coast of Africa on the western bank of the Red Sea. Their task was made easier by using the small cutting blades that they had chipped from a glass-like volcanic rock called obsidian.

To archaeologists, the tool and bone debris these humans left behind represents the oldest evidence on the planet of the beachcombing way of life. Crucially, their location puts them exactly at the jumping-off point from which many geneticists believe humans set off some 40,000 years later on a migration that quickly took them as far as southern Asia. Because sea levels were lower then their walk could have continued without a major sea crossing as far as modern-day Java. It could have taken their descendants less than 10,000 years to reach Australia even though their journey was disrupted by the greatest volcanic eruption ever experienced on earth.

The footprints they left in the sand have long since disappeared, but the route they took can be clearly traced in the distribution of the genes of their descendants across the modern world. This is the task of the phylogeographer: to track in time and space the path of human DNA as it mutated and spread around the world.

Phylogeography: the mapping of human migrations

For several decades geneticists have been working alongside archaeologists, palaeoanthropologists, climatologists and linguists in the new field of phylogeography: the mapping of the history of mankind's migrations. The two key questions their studies seek to answer are: where did the ancestors of modern humans come from, and how and when did mankind spread across the planet?

Research during the past two decades has broadly answered the first question. DNA analysis shows that there is more human (read genetic) diversity – the hallmark of an old and long-established population – in the average African village than has been found anywhere else in the world. It follows from this observation that the relative lack of genetic diversity in other parts of the world is the result of their being colonized by just a small number of the original diverse African population, a type of selection process that geneticists describe as a genetic bottleneck. Today the 'Out of Africa' thesis that explains these findings is almost universally accepted.

The genetic detective work that has pieced together the history of human migrations over the last 150,000 years has been built on analyzes of both maternal-line and paternal-line ancestries. While the earliest academic studies looked at the maternal-line inheritance and used mitochondrial DNA tests, more recent paternal-line Y-chromosome studies have added a great deal more detail to our overall understanding.

While the two lines of study reveal different aspects of the history of the spread of modern humans, they also corroborate each other. Some details are, of course, disputed among geneticists, and the attempt to reconcile the models of human development of last 150 millennia suggested by geneticists with the more established models held by many archaeologists can still prove contentious. For one thing, the absolute dating of genetic mutations is nowhere near as precise as the dating of events in climatology (using ice cores) or archaeology (using carbon dating). That said, the scientists aligning these different inputs and dating methods are increasingly revealing a consistent story of human migration and colonization.

The major factor that has stimulated human migration has always been climate change, principally through its effect on food supplies, sea levels, seawater salinity and vegetation, but also in specific instances where shifts in climate have literally opened geographical corridors through which mankind has then been able to walk to new lands.

While the migration history of our species Homo sapiens fits into the last 150,000 years, the remains of earlier species of human have been found – including one skull in Georgia in the Caucasus, 2,500 miles (4,000 km) from eastern Africa – that date back perhaps 1.7 million years. Such an early migration from Africa was once thought highly unlikely and was seen as evidence that Homo sapiens must have arisen independently in several places outside Africa.

The debate between a multi-regional thesis of mankind's origins versus the 'Out of Africa' model still rumbles on, but while the earth's climate has over the last 2 million years been generally been icier and drier than it is today, there have regularly been brief warm wetter spells – known as interglacial optima – that would have enabled regular migrations out of Africa. When the most recent interglacial optimum occurred just 8,000 years ago, the climate of Africa was so different that cave paintings found in the central Sahara depict the savannah animals we associate today with the plains of southern Africa.

Tracking our genes

Geneticists track both mitochondrial and Y-chromosome DNA in their studies, but as the focus in this book is on the Y-chromosome, for the rest of this chapter I'll use that strand to illustrate their findings.

The reconstruction of the network of human migrations using the Y-chromosome has been built around the results from a range of SNP markers. Geneticists use these on/off markers to work out which ones changed a long time ago and which ones changed, in genetic terms, relatively recently. The resulting branching diagram – a phylogenetic tree – demonstrates how the different groups descend from each other. Each branch in the tree – a clade or haplogroup – is differentiated from its ancestor generally by a single marker.

With the phylogenetic tree worked out and the clades classified, geneticists can then plot them on a map, starting with the original male ancestor of us all back in Africa, to highlight the migration patterns that would most likely have led to the present-day distribution of markers found across the world.

Would you Adam and Eve it?

While creatures that can be classified as humans emerged on the African continent more than 2 million years ago, only relatively recently did a group of anatomically modern humans arise in eastern Africa. Indeed their most genetically direct descendants – and our closest link to those original humans – are still found in Ethiopia, Sudan and parts of eastern and southern Africa. It's now understood that all humanity is descended from a man and a woman who were among the estimated 2,000 to 10,000 humans who were alive some 190,000 years ago in Africa. As far as can be calculated, however, these two people lived at widely different times.

Thus while the results reveal – by using the Y-chromosome tests – that all men can trace their paternal descent from a single man and that – by using the mitochondrial tests – everyone can trace their maternal line back to a single woman, it's important to understand that the results are revealing lines of genes, not the identities of specific ancestors.

These unique ancestors do not represent two original humans – regularly referred to as Adam and Eve – who lived at the same time and shared children together. Instead, they tell us that of all the humans alive at the time of the earliest known genetic mutation that we're testing for today, only one man and one woman produced descendants that are alive today to be tested. These two separate individuals were the progenitors of successful gene lines; no descendants survive of the other humans alive when they were alive. Defining where and when they lived is still much debated; creating fictional identities for them, as some writers are wont to do, is pure poetic licence and often quite confusing.

Climate change triggers human migrations

Today we live in an interglacial period when the global climate is warming. This follows the end of an Ice Age some 18,000 years ago when one-third of the earth's land mass was frozen. The extended Ice Age cycle lasts roughly 100,000 years and the last one contained several shorter cycles that produced mini-Ice Ages interspersed with relatively short periods of warmer weather – known as interstadials – none of which lasted for more than a few thousand years. Even within the historical time-frame of the last 2,000 years, during which the world has been getting

Table 1 Climate change and human migration

Climate	Africa	Arabia	Asia/Oceania	Europe	Americas
ICE AGE					
COLD					
INTER-GLACIAL HOT / WET	Failed migration to Near East				
COLD AND VARIABLE					
ICE AGE	Abdur beach-combers walk to Arabia		Arrival in South East Asia		
ICE AGE			Arrival in Australia		
ICE AGE				Near East migrants reach Europe	
ICE AGE					North Asian migrants in the Beringian refuge
ICE AGE			Central Asian migrants reach Europe		North America peopled
INTER-GLACIAL					

Scale on left: 140,000 BP · 120,000 BP · 100,000 BP · 80,000 BP · 60,000 BP · 40,000 BP · 20,000 BP · today

74,000 BP Toba eruption

32,000 BP The end of the Neanderthals

Data based on Oppenheimer

warmer, there have been periods of relative cold, for example the five centuries between 1350 and 1850. Within such modern time-frames it's even possible to identify social changes that have been stimulated by climate change, for example the abandonment by the Vikings of their Greenland colonies around 1450 as the climate deteriorated after the end of what historians now call 'the medieval warm period'.

Climate shifts have been triggering changes in human society for tens of thousands of years before the present era. From an African perspective, climate change has acted as a kind of switching mechanism that opens and closes the two 'Out of Africa' migration routes, the first that runs to the north through modern-day Egypt and into the rest of the Near East, the second that crosses from Eritrea to Yemen at the southern end of the Red Sea. Along the northern route climate change alternately created the Sahara – an effective barrier to migration during drier periods – or during wetter periods made the desert area more hospitable and thus opened it as a migration route to the east through to the Arabian peninsula and the rest of the Near East.

Around 120,000 years before the present (BP), during the last interglacial optimum and prior to the most recent mini-Ice Ages, a group of humans walked northwards through the grasslands of the Sahara to modern-day Egypt and Gaza. They survived there, however, only until around 90,000 years BP, most likely dying out during a prolonged freeze that turned the Mediterranean coastal lands of the Levant into an extreme desert and which closed shut their route back to Africa. The migration story that the rest of this chapter describes was thus not the first that humans had undertaken, but it was the first successful one.

The peopling of the modern world

Some 10 millennia later, around 80,000 years BP, as the mini-Ice Age that ultimately killed off the previous group of northern-route migrants was giving way to a brief warm interstadial period, another group of humans – most likely the direct descendants of the Abdur beachcombers whom we met at the beginning of the chapter – headed east, crossing from present-day Eritrea to Yemen where they began walking along the shores of southern Arabia and onwards to Iran and Pakistan. Their crossing would have been easier than today as sea levels were lower (as, following a colder period, a great deal of water was still locked up in the polar ice sheets). The Oxford scientist Dr Stephen Oppenheimer, in his book *Out of Eden*, suggests that it is possible that their migration was prompted by changes in the salinity in the Red Sea as sea levels rose and the resulting loss of their regular food sources.

As beachcombers, their descendants found that the easiest expansion route was simply to follow the coast of India and thence into Indo-China, the Malay peninsula and the archipelago of modern-day Indonesia. Moving from India to south-east Asia and southern China may have taken as little as 5,000 years and Oppenheimer's theory suggests that

Figure 3 Path of Y-chromosome DNA from Africa to Asia and Europe

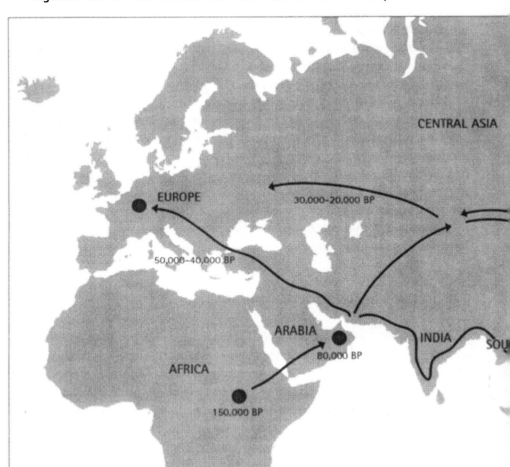

within a further 5,000 years, by 70,000 years BP, some descendants had reached Australia.

It is generally accepted that all people alive today are descendants of the tiny group of people who walked out of Africa 80 millennia ago. Much more contentious is the debate over which area of the Middle East – the options range from Turkey right down to Iran and Pakistan – was the main starting point from whence different subgroups of the initial African migrant group left to populate south-east Asia, Europe, Siberia, and ultimately the Americas.

Geneticists are still pretty much divided between those who favour the

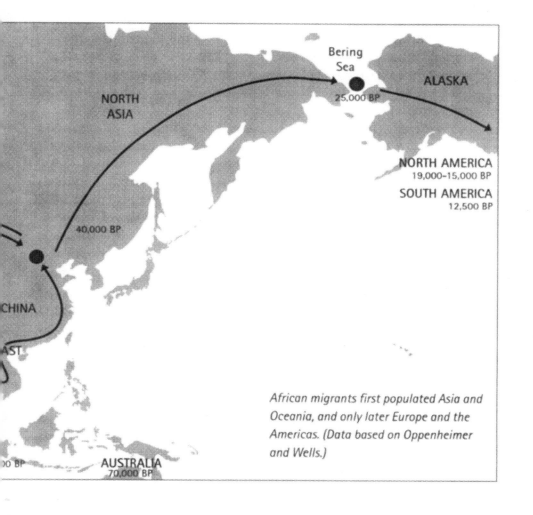

African migrants first populated Asia and Oceania, and only later Europe and the Americas. (Data based on Oppenheimer and Wells.)

Near East and those who favour the southern areas of the Middle East, but Oppenheimer makes a strong argument for the Iran/Pakistan border as the area where the out-of-Africa group's descendants split in three main directions.

The principal expansion route followed the coastal margins into Asia. As well as the coastal route I've just outlined, Oppenheimer identifies a second expansion pathway that left the coastal areas of modern-day Pakistan to follow the region's major river, the Indus, up into Central Asia. There these migrants met some of the descendants of the first coastal route group that had independently moved into China and east Asia and who by 40,000 years BP had doubled back and were expanding westwards along the route we know today as the Silk Road.

The third expansion pathway led from the Iran/Pakistan area westwards through modern-day Iraq and Turkey and directly into Europe via the Balkans, an area reached perhaps as early as 50,000 years BP. These migrants were joined perhaps some 30,000 years ago by people from central Asia, descendants of those who had taken the second expansion route.

The Americas were peopled by descendants of the second expansion route who had gone into central Asia, some of whom chose to head east towards eastern Siberia instead of west towards Europe. Around 25,000 years BP descendants from both the second and third expansion routes managed to cross the land bridge – the now flooded continent of Beringia which connected Siberia and modern-day Alaska – and moved into modern-day Canada and the USA. Once in the Americas their descendants moved overland, reaching the eastern seaboard of the continent possibly as early as 19,000 years BP, and also along the western coast to reach northern Chile no later than 12,500 years BP.

At the core of this global migration story is the identification of Pakistan and the southern Gulf area as the jumping-off point for these three post-African dispersals, a theory described by Oppenheimer as fitting the genetic evidence that places the oldest non-African gene lines in south and south-east Asia, not in Europe.

Underpinning the theory are data from a known event, the massive volcanic eruption that took place some 74,000 years ago at Toba in northern Sumatra which created an ash cloud so immense that it stifled India and has been recorded as far away as Greenland. Toba's ash deposits provide a globally dated calibration point that can be used as a reference mark in archaeological digs and climate samples. Amazingly, the effects of the Toba eruption appear to be visible even in the genetic data of modern Asia populations as the predominant genetic line that spread with the beachcombers from Africa to Australia has younger sub-groups that are found only in India. This finding implies that these sub-groups had expanded back into this geographical area, most plausibly when India once again became habitable after it recovered from the Toba ash falls.

Climate change also stimulated the peopling of Europe. Around 50,000 years BP, during a brief interstadial when the climate became wet-ter, a land migration route from the Gulf area opened up along a corri-dor from Iraq to Turkey known as the Fertile Crescent. This route allowed humans to reach both the eastern Mediterranean countries and the Caucasus region. It was this group that was to be the first to success-fully colonize Europe.

Since the last Ice Age

The defining climatic event for modern-day humanity was the expansion and retreat of the last Ice Age which is technically described as the Last Glacial Maximum (LGM) and more colloquially as 'the big freeze', which took place roughly 22,000–15,000 years BP. When it began, previ-ously habitable areas became covered by ice. This forced the human pop-ulations to migrate and cluster into smaller isolated communities, areas described as genetic refuges. In Europe these refuges were concentrated in three areas: the Basque country between France and Spain, the Balkans and the Ukraine. The reduction of the human population created a huge genetic bottleneck. When the Ice Age began to end at around 18,000 years BP those survivors remaining on the European land mass re-expanded northwards as the ice retreated.

In Asia the picture is less clear. While some humans lingered and sur-vived the Ice Age in southern Siberia, many others simply moved on as the ice advanced, most likely towards south-east Asia. As in Europe, they too later re-expanded back as the ice retreated northwards.

The debate over the peopling of the Americas could be a book in its own right, but the likeliest theory is that the descendants of several dif-ferent genetic groups from Asia had reached there well before the LGM.

Many mtDNA lineages found today in America appear to have arrived there before the ice physically closed the corridor between Asia and North America. As the last Ice Age took hold, descendants of some of these early migrant groups survived in a refuge in Beringia, a continent-sized land bridge that existed in the period 25,000–11,000 years BP and which covered an area that included modern-day Alaska and eastern Siberia. Survivors of this genetic bottleneck would have re-entered the Americas as the ice retreated around 15,000 years BP and before rising sea levels drowned the continent of Beringia some 4,000 years later.

Arguments remain

This chapter is a bare outline of a massive multi-disciplinary reconstruction exercise that is still contested in many details. Some of the contentious issues of the past are, however, now generally resolved. To start with, there is no evidence in modern-day human populations of any genetic intermixing of Homo sapiens and Homo neanderthalensis. It appears, too, that European Neanderthals and the south-east Asian members of our immediate ancestor species, Homo erectus, had all died out no later than 30,000 years ago.

As 15 mtDNA lines have been dated to origins in Africa at more than 80,000 years BP, the African continent is now accepted as the origin of the oldest human gene lines. The arguments over the peopling of Europe are, however, more persistent. While Oppenheimer's compelling thesis is that Europe was colonized through a westwards migration from the Gulf/Pakistan region no earlier than 50,000 years BP, other theories have yet to be laid to rest. Some geneticists variously believe Europe's ancestors could have made their way to the Mediterranean coast up the Nile in the period 120,000–90,000 years BP, that some could have survived as a refuge population in North Africa and Egypt during an earlier mini-Ice Age and re-expanded into Europe from there, or that the northern African route through Egypt was also viable in the period 45,000–20,000 years BP.

To show how fresh these debates are, research teams are currently working on an analysis of DNA samples from modern-day Ethiopians in a bid to date accurately the early African lines. The Americas are also an area of contention and some experts believe the continent was only peopled after the LGM and not before. Finally, Oppenheimer's thesis that Australia was peopled by 70,000 years BP is a challenge to orthodox opinion, built up around archaeological evidence, which places that event much later.

In many cases these arguments remain unresolved because of the difficulty of calibrating the different methods used in each discipline to date their evidence. These technical issues are being tackled even as you read this book. All the debates I've outlined are described in more detail on this book's website.

Classifying gene groups

The classifications used in phylogenetics and phylogeography are far harder to comprehend for a layman than the truths they seek to explain. Up until 2002, academic geneticists generally created their own classification system to explain their results. No fewer than seven such systems exist in academic papers written around that time. That year, however, a group of scientists known as the Y-Chromosome Consortium (YCC) under Dr Michael Hammer at the University of Arizona agreed a standardized classification system for describing the branching network of genetic clades revealed by the Y-chromosome. These clades are denoted with a capital initial letter and show subclades derived from them with a combination of lower-case letters and numerals; for example, R1a1b is a subclan of the main R clan. The phylogenetic chart of humankind shows how each clade fits together and the marker that distinguishes it from its ancestor clade or subclade; for example, R1a1b is different from other R1a1 subclans at the genetic marker labelled M157.

The most accurate way to label a clade is to combine the clade letter and the distinguishing marker. Needless to say, this is the one method that it appears is never used. First-time readers need a concordance to negotiate their way through these thickets, made all the more confusing by the use of fanciful monikers and clan head biographies intended to make science more consumable.

One other potential confusion is that the same initial letters can be used for both mtDNA and Y-chromosome clades, often without further distinguishing marks. In this book I've labelled them; if it's not labelled you can assume I'm talking about Y-chromosome clades.

The peopling of Europe

Far from being a pioneer in human development, Europe – geographically remote and burdened by a relatively inhospitable climate – has until recently been a recipient of cultural developments discovered elsewhere. Only the Americas were settled later. The genetic evidence broadly agrees with the main theories built up from the archaeological evidence which reveals that the spread of tools and agriculture across the continent during two pre-Ice Age east-to-west migrations coincided with key changes in tool technology. Europe was peopled in waves as new technologies were carried across the continent by peoples most recently of south Asian origin.

The first technology change occurred at around 46,000 years BP (known as the Aurignacian or Upper Palaeolithic period) while the second occurred during the period 30,000–21,000 years BP (known as the Gravettian or Middle Upper Palaeolithic period). The range of dates is partly indicative of the time the technology took to establish itself across the continent – Aurignacian tools, for example, took around 12,000 years to spread from modern-day Bulgaria in the Balkans to Portugal on the Atlantic seaboard – and partly because these changes are often difficult to date using traditional carbon-dating methods.

The key features of the mitochondrial DNA picture in Europe and the Near East were first laid out in a major study published in 2000 by Dr Martin Richards, then of Oxford University, now at Huddersfield University. The study identified 11 European mtDNA founder lines that had origins in the Near East, the oldest of which – mtDNA clan U5 – dates back some 50,000 years. This clade is one of four identified as the founding lines of the Near East; while it is found in Europe and parts of the Fertile Crescent it is not found in East Asia. Even more clearly, this subgroup U5a is common in the Basque country, an area identified as an Ice Age refuge for Europe's original population and preserver of some of its pre-Ice Age genetic diversity.

The second-oldest mtDNA clade – HV, the ancestor of both the mtDNA clans H and V – entered Europe through the Russian steppes and into modern-day Germany around 33,000 years ago. It is now the most common mtDNA clade in Europe. It seems to have arrived in Europe by a different route from the others, via north-west India and then to the Caucasus region or possibly via the Central Asian steppes further east before looping back to the Caucasus.

Figure 4 Phylogenetic chart of the main Y-chromosome clades

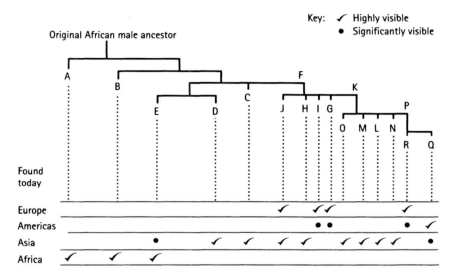

This simplified phylogenetic chart shows the modern-day distribution of the main haplogroups (clades). Note that the oldest clades – top left in the chart – are found in Africa while the youngest – bottom right – are more widely distributed. The full chart is divided into 131 subgroups and can be viewed on this book's website.

A global Y-chromosome study analogous to the Richards' mtDNA study was carried out by Dr Peter Underhill at Stanford University, California, also publishing its results in 2000. The branching structure of Y-chromosome clades that this team laid out forms the basis of the YCC's standardized haplogroup labels published in 2002. Figure 4 offers a simplified view without showing the markers associated with each clade. A full version can be viewed online.

An analysis of Y-chromosome DNA in Europe shows a similar picture to that revealed by the mtDNA data. For example, the male J clade, which is common in the Near East and progressively less so in Turkey, Greece and Italy, mimics to a degree the distribution of the mtDNA U5 clan. Other male genetic lines show concentrations in Finland and Russia (clade N), and eastern, central and western Europe (clade R). Clade R is, in fact, the modal clan worldwide and ancestor of half of all European males. It's the clan that my DNA results indicate that I belong to. Clade R's ancestor, clade P, is confined to Central Asia, India and America as well as being common in the British Isles and Basque regions. The fact that these are both on the extreme edges of Europe suggests that they represent an earlier arriving population.

Why is this relevant for genealogists?

This chapter has given just a broad-brush view of the history of human migration from Africa and the populating of the planet in the period 200,000–10,000 years BP. But what is the relevance of this to family historians who are generally only looking at the past 500 years, and at most two millennia?

Some of the early Y-chromosome test projects that involved genealogists in the period 1997–2002 did in fact use some of the SNP markers that are used to distinguish the different clades in the YCC classification. This was simply because the academic test labs knew the procedures for testing those markers and were still trying to identify the STRs that purpose-built genealogical Y-chromosome tests now use.

As of summer 2004, some commercial testing companies are offering tests on the markers that define the Y-chromosome clades while others are marketing tests for specific ancestral migrant groups. One testing company cleverly offers a comparison with known STR/SNP results and estimates your haplogroup for free. I'll consider these options in detail in Chapter 12.

If this investigation stimulates your interest, note that all the testing companies offer a mitochondrial DNA 'deep ancestry' test that will define the maternal ancestral line of anyone, male or female, who takes it. The results use the standard Cambridge Reference Sequence classification system so you will be able to benefit from future academic refinements to the overall mtDNA picture.

All the same, a man does not need to take a SNP test to learn something of his deep ancestral background; he can in fact learn a great deal of geographical information about his ancestry simply by taking the standard Y-chromosome STR tests offered to family historians. This geographical information can in turn reveal a great deal about where his direct male ancestors came from in the immediate period before surnames came into widespread use. We'll look at how to do this in Chapter 4, but before we look at how to interpret a Y-chromosome result we need to introduce another area that has a great impact on the overall analysis and understanding of a Y-chromosome DNA result: the surname.

More info at <www.DNAandFamilyHistory.com>

- Key migration routes from Africa to Europe, Asia, the Americas

- Diagram of the Y-Chromosome Consortium chart of clades

- Interesting academic papers on migration studies

Chapter summary

- All humans can trace their origins back to Africa, to a single man and to a single woman

- Europe was settled in several waves of migration from south Asia and the Caucasus, at the earliest around 50,000 years ago

- America was probably settled before the last Ice Age some 22,000 years ago by populations moving east from Asia

- Genealogists can learn a great deal about their 'deep ancestry' from the standard Y-chromosome DNA test designed for family historians

CHAPTER 3

A New Reading of Surnames

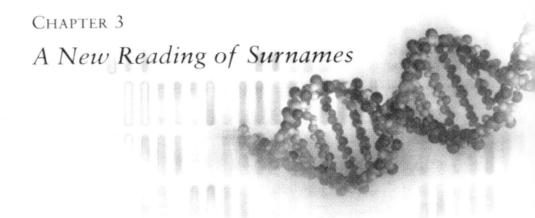

Genetics meets genealogy

As the use of DNA analysis in migration studies became established during the last two decades it didn't take long before several academic teams started to look for ways to expand genetic investigations into new areas. The impact of their resulting encounter with the community of family historians is still growing.

In Britain, archaeologists, historians and geneticists are well aware that the make-up of the modern population of these islands has been regularly refreshed during the past few thousand years by influxes of new genetic material arriving from continental Europe. The goal of identifying the dominant clades and typical haplotypes of the successive groups that have populated Britain since the last Ice Age was on the minds of several research teams at the end of the 1990s. While they chose two different routes to tackle the question, both routes quickly brought them into contact with the genealogical community with its focus on post-medieval history.

The first route that researchers took was basically a refinement of the standard phylogeographical migration study. The idea was to try to identify modern-day individuals who were more likely to represent a given group – for example, the Vikings – than would a randomly sampled 'average' Briton. The starting point was the expectation that, since invaders generally came from the east, the DNA signatures associated with earlier settlers are more likely to be found even today in the westernmost areas of the British Isles. Identifying the frequency of different haplotypes across the country should therefore reveal clear regional variations in west/east geographical distribution and hence allow haplotypes to be labelled as typical of an invader or of an earlier inhabitant.

In 1998, researchers in Dr Mark Thomas's team at University College London (UCL) began a project to look at the DNA signatures of the descendants of families with long pedigrees in the counties of Suffolk, Norfolk, Devon and Cornwall. By testing men who could trace their family history back to the sixteenth century, the UCL researchers expected to be able to pinpoint the differences in the relative frequencies of different DNA signatures between areas in the far west and the east coast of England. These results could then be compared not only against each other but against databases of samples of men from across the rest of the British Isles.

While the results of this original UCL project have still not been published, this type of regional study has since been carried out by many different teams sampling populations in England, Wales, Cornwall, Ireland and the outer Scottish islands as well as targeting geographical areas where specific groups were known to have settled, e.g. coastal areas which had major Viking settlements. Seen together, their results have started to create a picture of clade and haplotype distribution by region across the British Isles.

The second investigation route focused on the use of surnames. Geneticists have been predisposed to view surnames as reliable carriers of genetic information ever since the publication in 1985 of an influential book by the anthropologist Gabriel Ward Lasker titled *Surnames and Genetic Structure*. Lasker investigated how surnames could be used to estimate gene frequencies in the general population, so it was a natural step subsequently for others to investigate the way genetic material is transmitted within surname groups.

Working at the same time as the UCL team, Professor Bryan Sykes at Oxford University pioneered this route and it was the publication of the results of the study of his own surname, in the journal of the American Society of Human Genetics in the year 2000, which signalled the launch of surname-based DNA testing in its own right.

Several other academics and businessmen around the planet had simultaneously spotted the potential that DNA testing held out for family historians, thus creating a very interesting landscape for genealogists within a few short years.

The Sykes surname study's results

Many surnames derive from a specific location. Many more are associated for linguistic or historical reasons with one region in Britain rather than another – for example, the professor's surname is most strongly associated with the northern English county of Yorkshire – so using surnames to select men with a stronger than average likelihood of representing a given set of related genes is a logical choice.

The long-time view of historians of surnames has been – at least until recently – that a common name like Sykes must have been taken up as a surname by many unrelated people over an extended period of time rather than stemming from a single ancestor who adopted it at a specific time and place. It had been assumed that the word 'syke', which is often used to describe streams, springs and boundary ditches in Yorkshire, would have been adopted by many different, unrelated men across the county during the period when surnames came into use. Instead the professor's DNA study concluded that all modern-day Englishmen with the surname Sykes generally share a single common ancestor. What's more, that common ancestor might even have lived within the seven-century time-frame since surnames came into use. As the professor later summarized it, there may well have been 'an original Mr Sykes'.

I remember reading a summary of these results on the BBC News website shortly after they were published and wondering whether other family historians were puzzling over them as I was. My own research into the Pomeroy surname had led me to expect that our modern name-bearers are descendants of many different Pomeroy ancestors who had taken the surname independently of each other. The combined numbers of Pomeroy, Pomroy and Pomery name-bearers in Britain today is no more than a quarter of the numbers bearing the name Sykes. So how could a name as common as Sykes, with around 10,000 name-bearers living in England and Wales today, have a single common ancestor who lived as recently as seven hundred years ago?

While the original Sykes study raises many questions, there is no doubt that it popularized the use of DNA testing by family historians and began the process of turning genetic identification into a mainstream activity for genealogists. Since then the number of surname-based DNA studies has flourished, rising from zero to around 1,500 in just five years. As their results are published it is becoming clear that multiple-ancestor profiles seem to be less rare among surnames of English origin than the Sykes study's results might at first glance suggest.

The analysis of these surname-based DNA studies has, however, only just begun. But to understand how important it is going to become for family historians we need, before we go any further, to understand how surnames came into use.

How surnames were adopted

The adoption of hereditary surnames in England, Wales, Scotland and Ireland was neither legislated nor an overnight social change. It happened gradually over many decades, starting some seven or eight centuries ago in the south of England and finally in Wales overcoming the old patronymic system as recently as the nineteenth century.

In Britain, the idea of hereditary surnames originated with the Anglo-French nobility some three or four generations after the Norman Conquest in 1066. The senior line of a major family might in its title record its link with an original settlement in Normandy, while junior lines of the family used other multiple bynames based on a place, status or other attribute. The practice was widely established across England by end of the 1300s and by the time of the 1377–81 poll tax most of the ordinary townsfolk listed as taxpayers in England had a surname or byname.

While there's no evidence to explain how surnames become hereditary or why women began adopting their husband's surname on marriage, one factor may have been the growing need to assert one's identity and lineage, given that the amount of documentation in business and legal life was growing fast. Another reason might be that the post-Norman period was very uninventive in its choice of Christian names. A leading expert on British surnames, Professor David Hey, cites a 1379 poll tax record for part of Yorkshire where more than half of all men were called either John or William, a situation hardly conducive to clarity.

How surnames proliferate

The growth of documents introduces another interesting issue. Fixed spellings were not the norm in medieval records, indeed the spelling of a person's surname often varied even within the same document. For scribes, consistent spelling was not a virtue that needed to be practised. English is not a phonetic language so spelling variations multiplied as scribes recorded what they heard, there being no single agreed spelling to refer to. Surname orthography really only began to become fixed in the past two centuries, and firmly so only within the last hundred years.

To illustrate with my own surname, in Britain Pomery exists alongside the key variants of Pomroy and Pomeroy. Several things have happened to create this modern-day situation. Firstly, dozens of other variant spellings, such as Pomeray, Pomory and Pumry, have disappeared as they have been standardized into the three main surviving surnames. Secondly, these three surnames often cluster into specific geographical areas – for example, the spelling Pomery was traditionally found in Cornwall rather than in neighbouring Devon where Pomeroy was the norm. Thirdly, spellings were often consistent within a documented family, but had a tendency to change when family members moved away from their home parish or even, in medieval times, when a new priest took charge of the parish registers.

From this example one might expect that the total number of native surnames in use in Britain is likely to have declined over the years, and one would infer therefore that the number of people bearing any particular surname must, on average, be growing simply because the national population has risen ten times since surnames first came into fashion. At the same time, it has been observed that while most people in Britain today have a common or relatively common surname, most surnames are themselves very rare.

In fact, rare surnames turn out to be very common indeed. In *Family Names and Family History*, where many of the examples cited in this book are developed further, Professor David Hey recounts an analysis he made of the 1986 telephone directory for Sheffield, a city in the north of England, covering just under a quarter of a million private subscribers. He counted 19,510 surnames of which 8,318 (43%) occurred only once while 13,569 (70%) appeared only four or fewer times. Those 70% of surnames, however, accounted for just 21,376 – or 9% – of all subscribers. In other words, 30% of all surnames accounted for 217,180 – or 91% – of subscribers. While such a long 'tail' of rare surnames might be seen as specific to a small locality – perhaps other name-bearers were hiding in neighbouring areas – this kind of pattern is repeated at national level and also in other countries.

Observations suggest that, at least geographically, surnames are relatively stable. Many rare surnames are still found congregated close to where they were founded and even common surnames generally can be identified within a geographical area of likely origin. Some studies indicate that for many surnames the modern distribution is similar to that existing a century ago. Clearly there's a great deal of continuity at work here.

The study of surnames, as with genetic genealogy, is changing rapidly as large-scale computerized data sets, such as national censuses, tax lists and modern electoral rolls, become available. If the problem with surnames is that at the moment we know surprisingly little about them, genetic genealogy may well be the tool that improves this situation. To start to appreciate this we need to now consider the ways in which British surnames can be categorized.

Locative surnames

Surnames based upon a specific location are by far the most common type of surname in Britain. In many cases the location is as specific as an individual farmstead or a place no longer found on any map, in others a more generalized location such as a village or town. Some of these surnames have been identified because hundreds of years later the surname holders remain concentrated close to their point of origin and their family history has been traced back through early medieval documents.

Some of these surnames could conceivably have arisen because an ancestor travelled away from his home region during the period of surname formation, so the principal modern-day heartland of the family could therefore be in that new settlement area rather than the area of origin. Studies have shown, however, that overall more than half the surnames in Britain still have a statistically significant association with a particular locality.

Locative surnames are relatively more common in the parts of Britain that are sparsely populated with scattered farmhouses rather than larger settlements, for example in the far north and the south-west of England.

Like all surnames, locative names can 'mutate' when name bearers change their locality, their name then being pronounced differently and leading over time to a change of spelling. Some modern surnames are unrecognizable abbreviations of familiar places. The historian Percy

Hide Reaney, author of the standard work *A Dictionary of English Surnames*, cites the surname 'Deadman' as a contraction of 'Debenham', a village in Suffolk, as an example.

Given that locative names are so specific in their origin, logic suggests that many of these surnames will probably have a single common ancestor at their head. One would thus expect that DNA tests on modern-day name-bearers would reveal that they have identical or very similar DNA signatures.

Topographical surnames

The second common type of surname is that derived from a topographical feature in the landscape. These arose in the same way as locative surnames, though because they tend to describe a general feature rather than a specific location – e.g. Brook, Green, Hill, Wood – it is generally assumed that they must have arisen in many places across the country independently of each other. Philologists have shown that many topological surnames are recognizable by a prefix – e.g. Atwater, Bywater or Underhill – and sometimes by a suffix – e.g. Downer, which is found in Sussex, the southern English county that's home to the rolling hills of the South Downs.

One of the interesting recent developments in surname studies, stimulated by Dr George Redmonds, author of *Surnames and Genealogy*, is that some topographical surnames have recently been shown to be locative, i.e. they are linked to a specific location rather than to a general landscape feature. The relatively common surname Sykes falls into this category.

Given that topographical surnames are likely to have arisen in many different places across the country one would expect that the DNA results of men bearing one of these generic surnames would reveal its multiple ancestor origin.

Occupational surnames

Few of us would hesitate to conclude that surnames derived from occupations, the third main group of surname types, have anything other than multiple ancestor origins. Occupations were used as bynames long before surnames became fixed and hereditary. Every community would have required at least one person to have these skills, so within a parochial context occupation would have been a significant identifier. The commonest names – Smith, Wright, Taylor, Turner – are derived from occupations that were widespread in medieval times even though many surnames with occupational origins relate to now long-dead skills.

Some occupational names exist in regional dialect versions that have in turn produced distinct regional surname patterns – e.g. a fuller in the cloth trade was known as Tucker in the south west, Walker in the north and Bowker in parts of south-east Lancashire. A very few may be so regionally specific as to be of single ancestor origin.

Included in this overall category are status surnames which relate to office holders in medieval society. While some of these are recognizable to us today, e.g. Marshall, Chamberlain, many often are not, e.g. Reeve (from which we take the word sheriff) and Vavasour (a feudal tenant below the Norman rank of baron). As with topographical names, one would expect that the DNA results of men bearing one of these generic-type surnames would reveal that it generally has a multiple-ancestor origin.

Patronymic surnames

Patronymic surnames are derived, either directly or as diminutives, from personal names. The tiny number of forenames used in the Norman period led to an indulgence of diminutives with suffixes such as -cock, -kin, -kins, -et, -ot, -mot, -on, and -in. A son of Richard could end up with the surname Richards, Richardson, Dickson, Hitchcock, Higgins, Ritson, or even Dixon. Many old personal names generally survive in daily use only as surnames – for example, the Norman names Baldwin and Eustace. Such examples exist also for Saxon, Norse, Norman, French and German forenames.

Some surnames of this type give clues as to their origin. The suffix of '-s' in Richards is more commonly found in southern England and the Midlands while the '-son' in Richardson is more common in northern England and Lowland Scotland.

Some female names have also been transmuted into surnames. Professor David Hey cites the surname Marriott derived from Mary while George Redmonds cites Dyson as derived from 'Dy's son'.

One would expect that the modern-day bearers of common surnames like these would have had multiple ancestors, while those surnames that are rare today could easily stem from a single ancestor. DNA testing might therefore reveal either single- or multiple-ancestor profiles for patronymic-origin surnames.

Nicknames

The fifth type of surname is one derived from a nickname. The main examples of this type are high-caste titles – such as Knight, King, Bishop or Cardinal – and those based on a physical human characteristic – such as Wildgoose, Rust or Hurlbatt.

Many of the latter are tightly distributed in specific regions in the modern period, which suggests that they have probably stemmed from a single ancestor. The former group could just as easily have arisen in many different places at different times, so one would expect the DNA profile of these surnames to reveal a mix of single-origin and multiple-origin ancestries.

The role of genetic testing

The study of the origin of surnames has traditionally been the preserve of philologists, experts in ancient languages who created dictionaries of surnames by using the same methods to identify and classify them as already used to classify place names. If the historians have been unable to define what factors distinguish all single-origin surnames as a group from multiple-ancestor surnames, they do know how to find the answer: research the genealogy of each individual surname. Professor David Hey notes that local records regularly suggest more convincing etymologies than the dictionaries and George Redmonds has demonstrated in his Yorkshire studies that local genealogical research is the key to the verification of every surname's origins.

When discussing whether a family surname could have had a single ancestor, Hey has set out a rule of thumb that if the deaths of 50 name-bearers were registered in England and Wales in the five-year period 1842–6 then that average of 10 deaths per year suggests that the surname is small enough that its bearers could have had a single ancestor. In other words, our best expert's finding is a rule of thumb that differentiates between single and multiple ancestors based on a single sampling point. Can DNA testing improve on this?

The interesting question of whether a surname has a single- or multiple-ancestor origin is one that DNA testing has thrown into sharp relief, and while we've yet to reach firm conclusions that will tell us how to predict the outcome for individual surnames, it's clear that the growing corpus of DNA test results is going to shed light in this core area of genealogical research.

My own feeling is that this is not going to be such a simple task. The original Sykes study asserts that the relatively common surname Sykes has a single ancestor, findings that fit the historical and genealogical data researched by George Redmonds. As a result we can see now how this surname, once thought of as topographic, can be redefined as locative. However to derive a general rule from the pattern of the Sykes DNA results seems to me to be premature. The lesson from the historians and philologists is that you can't always tell by the name. The answer for each surname is only going to be resolved by looking at the precise genealogy of the surname in question.

One factor that will need to be taken into account is the effect of the Black Death in fourteenth-century England, which created a major genetic bottleneck at precisely the time surnames were being adopted.

History is such a developing subject that surprises occur regularly. It used to be a commonplace that geographical mobility was a feature purely of the modern era; after all, how could feudal labourers in medieval societies migrate away from their own parish? That was until a now-classic paper published 40-odd years ago demonstrated that movement and migration across parish boundaries was in fact the norm in pre-industrial Britain. Most people who left their parish nonetheless remained within their local 'country' – an area centred around the nearest market town – but the many who returned later to their home parish to take up their inheritance allowed the core families and surnames to remain associated with their locality despite this high individual mobility.

It's possible then that DNA testing will reveal that many more surnames have a single-ancestor origin than one might at first suspect. However, reviewing the five types of surnames I've outlined in this chapter, it seems reasonable to expect that most surnames in Britain are more likely to have multiple-ancestor origins than not. Topographical, occupational and patronymic surnames generally seem highly likely to have multiple origins while some locative surnames and those derived from nicknames could well have them too. Only rare nickname and patronymic-based surnames look certain to have stemmed from a single ancestor, while some rare locative surnames might also have done so. However, if half of all surnames have a strong location-specific element it's possible they will turn out to be locative by nature and have a single ancestor origin.

As I will show in Chapter 7, DNA tests are helping many family historians to advance their own family's genealogical research and to identify their surname's type and origin. Genetic testing is, however, simply one tool in the historian's armoury and it will take many more years and cross-referencing of DNA, historical and genealogical data before we can give even partially quantified answers to the major questions of how, for example, surnames are formed and survive. Even so it's clear already that DNA testing is going to improve radically our understanding of these processes and that the origin of the surname should be a principle focus of any genealogist's DNA study.

Now that we've surveyed the two main contexts for genealogical DNA studies – ancestral human migrations and the development of surnames – it's time to look in Part II at the ways that genetics specifically helps genealogists. In the next chapter I'll describe the different types of genetic study that family historians are organizing and how anyone can make the most of their individual DNA test result.

More info at <www.DNAandFamilyHistory.com>

- Academic DNA studies on UK populations

- The Sykes surname study

- Developments in the study of surnames

Chapter summary

- British surnames can be classified broadly into five types: locative, topographic, occupational, patronymic and nickname-based

- There are no quantified rules for determining whether a particular surname originated from a single ancestor or from multiple ancestors

- The trend in genealogical research, backed up by early DNA studies, is that single-ancestor surname origins may be more common that one might expect

- DNA testing has the potential to radically improve our understanding of how to classify surnames and how they are formed and survive

PART II

HOW GENETICS HELPS GENEALOGISTS

4 *Reading Your Ancestral Message*

5 *Y-Chromosome Study Scenarios*

6 *Single-Ancestor Studies*

7 *Surname Studies*

8 *Clan and Caste Studies*

CHAPTER 4

Reading Your Ancestral Message

Reading a Y-chromosome DNA test result

The testing companies will present your Y-chromosome result to you in a simple numerical format, as in Table 2.

Table 2 Sample 10-marker Y-chromosome result									
DYS no.: 19	388	389i	389ii	390	391	392	393	425	426
Result: 14	12	13	29	24	10	13	13	12	12

The numbers against each marker represent the allele value, the number of base pair (tandem) repeats counted at that particular locus or marker in the DNA sample.

Almost all the markers used in genealogical DNA tests are short tandem repeat markers (STRs). With this kind of marker every result will fall within a known range of results. This range will almost always have a very clear modal value – the most popular result – and the three most commonly found values will account for almost all the results that have ever been observed.

Looking at an individual marker's allele value doesn't tell you very much. Looking at a large number of markers is, however, very revealing. The value of a DNA test lies in the opportunity to compare your DNA signature with other people's. Even if your result for a single specific marker, or indeed for most markers, is the most popular result, if you test a large enough number of markers your DNA signature (or haplotype) will begin to look different to almost everybody else's. Bear in mind, though, that even if your set of results match with someone else's, this is only geneologically significant if you share the same surname. If you have different, unrelated surnames, there will be little point in searching for a genealogical link.

High-resolution tests versus low-resolution tests

Over the past few years it has become clear that genealogists are looking for a high degree of certainty when we try to decide whether two men's DNA results suggest that they belong to the same family tree or not. We're looking for DNA tests that offer a high degree of resolution during the matching analysis.

The first generation of DNA tests for genealogists tested 9, 10 or 12 markers. This number of markers is, however, now defined as a low-resolution test. Even though just a few years ago it was state of the art, today a low-resolution test can only be recommended in a few very specific circumstances, which I'll mention in Chapter 7 when we come to investigate how surname studies use DNA testing. It is, of course, still worth suggesting a low-resolution test when introducing people to the idea of DNA testing as it is relatively inexpensive and can often be upgraded later to a more advanced resolution test if the participant wishes to do so.

The highest-resolution Y-chromosome tests on offer in mid-2004 are 37-marker and 43-marker tests. This number of markers generates a DNA signature that is exceptionally able to highlight similarity and difference between test participants.

In many cases, medium-resolution tests that use 24–25 markers will work as well as the high-resolution tests, and for many people a medium-resolution test offers good value for money. What tends to happen, though, in any surname study is that two participants, unconnected or yet to connect their trees, who share the same surname and have an identical medium-resolution test result will immediately ask themselves, 'Maybe if we tested some more markers we'd be able to prove that we really don't share the common ancestor that the surname we hold in common suggests we might?' At this point it really seems like a good idea to invest a bit more money and upgrade to a high-resolution test in order to try to answer that new question.

Comparing Y-chromosome results

Along with the numerical results, the testing companies will send you a brief explanation based upon comparisons they are able to make with other samples in the databases of results they have access to. Some companies are able to tell you how many other people in their database of clients have an identical or close match with your haplotype. They may also have a mechanism that allows you to contact them. Furthermore, if you have already enrolled in a DNA surname study that is registered with your testing company you can expect that your study organizer will be able to give you some more details about how well your DNA signature matches the other members of your surname group.

Even without that help, on your own initiative you can start to interpret your results yourself by using four online databases of Y-chromosome results, several of which have been designed by the testing companies with surname DNA studies in mind. In most of these databases you can search for matches by surname as well as by DNA haplotype, though the basic family tree data linked to the results are generally only useful as a rough guide. In some databases you can email other people who have submitted their results while in others the results are anonymous. One database that holds anonymous data does, however, present the overall geographical distribution of haplotypes based on where the DNA samples were taken in Europe, Asia and America.

Using the free-to-use online databases

The most important thing to realize about all the free online databases of DNA results is that they will not provide you with definitive answers. They are, however, extremely useful at giving you additional data, which you can use as background for your genealogical research and as pointers for your own investigations. Perhaps the best way to think of them is that they provide you with a context for your DNA results.

One word of warning. In the databases that allow us, as owners of our test results, to submit our own data to the database, some of those data will be inaccurate. Any database where there is no process to verify the results entered in them will inevitably collect bad data. This doesn't make them useless, but it does mean that you have to be wary of unusual results. Databases that are made up of results derived from academic test programmes are in contrast highly reliable.

Table 3 Public databases of Y-chromosome test results			
Name	**Web address**	**Type of result data**	**Approx. no. of records**
YBase	<http://ybase.org>	Self-submitted	2,500
YSearch	<www.ysearch.org>	Self-submitted	6,300
SMGD	<http://smgf.org:8081/pubgen/site34.jsp>	Academic/Lab	8,700
YHRD	<www.yhrd.org >	Academic	25,000

Note: SMGD, Sorenson Molecular Genealogy Database; YHRD, Y-Chromosome Haplotype Reference Database. Figures as of mid-2004.

I'll use the set of Y-chromosome results laid out in Table 2 (see p. 50) to show how you can use these databases to analyze your results.

The YBase database

The YBase database is run by a firm called DNA Heritage, based in Dorset in England. In June 2004 it contained 2,342 results, by my estimate about 10% of all the commercial Y-chromosome tests done worldwide at that date. You can check any haplotype without first entering your own results.

Start by checking – see Alphabetical List – whether anyone with your surname has submitted their results; this could alert you that there may be an organizer of a surname study for your name.

Secondly, check whether anyone else has your DNA signature – see Search Haplotype. The database allows you to enter individual marker results for up to 36 of 49 different markers and to search for matches on as few as 8 markers. By entering the data for my 10 markers I can retrieve all the other results that are an exact match, i.e. we share the same haplotype at the low resolution of 10 markers. This shows me that 10 results in the database match mine on all 10 markers. The surnames recorded against those matches are Johnson (four counts), Carden, Clarke, Cotten, McGregor, Staples and Thomson. All these names sound British and none looks like a highly rare name, so as I'm British too there's nothing eye-catching there.

Thirdly, look at the statistics that DNA Heritage has derived by pooling all of the results in the database – see Statistics. These show the distribution of the individual allele values in the entire database. Checking my result values against each marker in turn I see that in 9 out of 10 instances my result is the modal value found in the database; clearly my genes are pretty common! The only exception is DYS 391 where my result of 10 is the second most-common result, not the modal.

Fourthly, at the foot of the same page have a look at a feature that tries to predict to which haplogroup my haplotype results suggest that I might belong. Few people have taken a haplogroup test and this table appears to be based upon results observed from only about 750 people. My closest match appears to be with the haplogroup clade R1b (from Chapter 3 you'll recall that the R clade is the ancestral line of half of all European males) except for my result at DYS 391 of 10 which shows a potential association with the R1a haplogroup. I conclude from this that the database is not yet large enough to distinguish precisely my haplogroup at this low resolution.

The YSearch database

YSearch is funded by a testing company called Family Tree DNA. In June 2004 it contained 5,607 records, by my estimate about 20% of all the commercial Y-chromosome tests done worldwide. Unlike the YBase database, in order to view matching DNA signatures you must first enter your own Y-chromosome results. However, once you've done this there are a wide range of useful search options. To illustrate some of these you can view the matches found using my own results.

Start by clicking on the tab – Search For Genetic Matches – and enter my result ID code – UQUQV. If you select matches on 8 markers the database returns 254 identical matches. If you select 10 markers it reveals 6 exact matches with men bearing the surnames Johnson, Carden, Cotten, McGregor, Staples and Guinan.

Next, tick the box on each result line in the column 'Compare', then click the linked word 'Compare', and then click the option – Show comparative Y-DNA results. This produces a table that reveals the full DNA signatures of all seven of us, including additional markers that were not part of my original database search. If I click instead the option – Show Genetic Distance Report – this also shows me the testing company for each person used.

At this higher resolution I can clearly see that no two of us have exactly the same results. So although the low-resolution 10-marker DNA signature search process has found some matches, this tells me that the six other men who appear to match me at a low resolution do not match each other at a higher resolution. From that I can guess that, even if I knew my allele values for a greater number of markers, only one of the other six men might then match me. This knowledge could well stimulate me to ask the lab to test some additional markers for me so that I can find out which of them is in reality the closest DNA match to me, though from a genealogist's perspective this is not particularly useful information as we don't share a surname.

YSearch has the facility for registered users to upload their family trees in Gedcom format.

Sorenson Molecular Genealogy Database

The Sorenson Molecular Genealogy Database is run by the Sorenson Foundation, which is in turn linked to the Sorenson Group of genetics companies, which includes the genealogical DNA testing firm Relative Genetics. The database is partly populated with Y-chromosome data taken during a global sampling exercise started by Brigham Young University in the year 2000 but since 2003 now part of the Sorenson Foundation. In June 2004 it contained 8,735 records. While you can search the database by marker you cannot submit your own results.

This database has the potential to be an extremely useful tool for genealogists. However, it is presently marred by a poor user interface which makes it difficult to use. The haplotype search results show whether a match has been found in the database and then displays a thin pedigree chart that the participant submitted when they gave their DNA sample. However, the likelihood of finding a genealogical connection by this method is extremely remote. Worse, you can check to see if a surname is found in the database but you can't search by surname. Not only is it not clear how many records there are per surname in the database, if your DNA result is not an exact match you won't ever see the results of people who share your surname.

The goal of the Sorenson Foundation is to create the world's largest database of DNA results so that it can start to analyze the DNA in the other 22 chromosomes, i.e. DNA that is not on the Y-chromosome. What we see now is a first draft of something that may become clearer in due course.

Y-Chromosome Haplotype Reference Database

The Y-Chromosome Haplotype Reference Database (YHRD, formerly known as the Y-STR database), is run on different principles. Organized by a committee of European academics, it is the central repository of Y-chromosome DNA results submitted by members of the International Forensic Y-User Group. Their standard set of nine markers is also used to calibrate the test procedures of forensic labs worldwide. Eight of these markers are used in the genealogical DNA tests now offered by commercial labs, thus making this a very useful low-resolution reference tool.

The database can be sampled on a worldwide basis (25,066 records in July 2004) or separately for Europe, Asia and the Americas. There is no surname or ancestral data linked to this data. The results simply show the number of men tested in each of dozens of locations per continent that match on the markers that you have inputted to check. You can input data for all nine markers or as few as one marker. (As this book was going to press this database was in the process of relaunching online and replacing its previous version at <http://ystr.charite.de>. The remaining data has been taken from the YSTR database.)

Using my sample DNA signature I have data that allows me to search on 7 of the 9 markers in the database. Checking the Europe option the search finds 350 matches, around 2.5% of all samples or roughly 1 in 40 in that segment of the global database.

The real benefit of this database is the additional geographical information it will generate for any haplotype checked. Click on – Geographical Information – to pull up a map showing in red the locations where this DNA signature was found and in blue where it has not been found. This map is exceptionally useful for rare haplotypes as it can pinpoint with clarity the specific cluster of locations where it is primarily found today. But even for a common DNA signature such as mine it can generate a wealth of detail.

The next step is to cross-reference the map with the box marked – Population Query Summary – below it. This table records the number of men with this DNA signature in each location, so it is possible to work out quite quickly in which cities and regions the highest percentage of matched results are found. For my sample this appears to be southern Ireland, Portugal, northern Spain, the Basque region and Germany. My result appears to be relatively uncommon in southern Spain, Scandinavia and eastern Europe, and it is not found at all in Finland, Romania, Bulgaria and southern Italy.

My result therefore appears to be strongly weighted towards the western part of the continent, and if we accept that Europe was populated by successive waves of migration from east to west then this could well indicate that my haplotype links me with the relatively earlier-established population.

A word of warning

Do treat the information derived from the two self-submitted results' databases (YBase and YSearch) with a degree of caution. Just to repeat, in any database where there is no quality control or data entry checking process you can be sure that some data – perhaps as much as 2% – will be mis-entered and thus become examples of the dreaded 'clerical mutation'. YSearch, to be fair, partly overcomes this problem by allowing men who tested with Family Tree DNA, the sponsors of the database, to submit their results automatically without rekeying them.

Secondly, from time to time the test labs undertake a recalibration exercise to check that their results are comparable with results produced by other labs. If they decide that they need to adjust their reporting process they may retrospectively reconfigure all the results on a particular marker for all of their own clients. Such changes are easy for the testing companies to make within their own databases of client results but are unlikely to be passed through to third-party or external databases. As the difference is usually only a single value, say from 13 to 14, it's usually impossible to spot this type of problem when looking at the raw data.

Thirdly, some people entering their data in the self-submitted results' databases will inevitably have missed or misunderstood the instructions for each database that tell submitters how to adjust the raw test scores that they have received from the test companies to make them compatible with that database. There are a growing number of these adjustments.

These comments are not meant to deny the usefulness of the online databases, merely to remind users that the first duty of a genealogist, as with documentary research, is to remain sceptical and to check everything. I maintain a list of these problems and further details about each database on the website accompanying this book.

New knowledge learned

The low-resolution DNA signature that I have used for this exercise has 10 markers, less than a quarter of the number of markers used in the highest-resolution tests that are now the benchmark for genealogists. Even so, by using these four free-to-view online databases I've been able to generate a lot more information that is useful to me as an individual and which will help me put my results in context if I take part in a full-blown surname study.

To summarize, it looks as though my DNA could be consistent with that of one of the pre-Ice Age migrant populations that ended up in a refuge in the Basque region or an early resettlement after it. In technical terms my haplogroup will probably turn out to be a subgroup of R1b. Also, while several men with British surnames have the same low-resolution DNA signature as mine, no one with a name remotely likely to be a variant spelling of my own surname has yet submitted their results in these public databases. On a low-resolution basis my haplotype is not very rare in Europe.

Clearly it will be worth my while to return in, say, a year's time to see if additional results added to these databases can create a clearer picture.

Types of mutation

At this point it is time to consider one of the trickiest concepts in DNA result analysis: determining when a match is a match. Put another way, just how similar must two DNA haplotypes be before we can say that they are not different?

On one level this is an issue about the level of resolution, or how many markers, are being compared. But one needs to understand at the same time that while a positive match on all markers is always a very useful indicator of similarity, the absence of a positive match can be information that is almost as useful for a genealogist.

The best way to approach this issue is to look at the two different types of mutation that can occur at the markers used in genealogical DNA tests: the single-step mutation and the double-step mutation.

Most mutational changes result in just a single incremental difference in a marker's allele value – for example, a single-step rise in the number of base pair repeats found at that marker from 15 to 16 or single-step fall from 15 to 14. Note that there is no significance in the direction of the change, only in the fact that a change occurs. In some cases a single mutation event may result in a double-step change, for example, from 15 to 17 or from 15 to 13. The statisticians who work on genetics projects have some figures built into their calculations that take account of this, but from our point of view the important fact is that double-step changes appear to happen even more rarely than single-step changes.

Matters are made more complicated because any marker can over time mutate in either or both directions – for example, a marker could mutate through a single-step rise from 15 to 16 and subsequently mutate through two single-step falls back from 16 to 15 and then to 14, the overall effect exactly resembling a one-time single-step fall from 15 to 14. This may not seem significant from the genealogists' point of view, but it's good to remember that no one is currently exactly sure how often the different types of change actually occur.

As the allele values found at each marker on the Y-chromosome tend to change through mutation only rarely, a single difference in the allele value at a single marker between two haplotypes – a single-step difference – may not indicate a huge genetic difference between two men who share the same surname. It could just be that one of them, or one of their ancestors other than their shared common ancestor, took on that single mutation when they were born. It's even conceivable, but not at all likely, that this kind of mutation might have happened more than once within their respective branches of their shared family tree so that their haplotype results appear different at two markers.

Rates of mutation

There are three key factors to remember when considering mutations and comparing Y-chromosome results. Firstly, some markers mutate more often than others, so if two results are different on two slow-to-mutate markers this change has probably taken more generations to occur that if the results are different on two fast-mutating markers. Secondly, haplotypes are not uniformly spread around the world. In Europe, some haplotypes are common whilst others are highly rare. Clearly it's easier to analyze a rare haplotype, but we live in hope that the common haplotypes that most of us bear will become better known in future. Thirdly, we all can see that a comparison between two haplotypes using 12 markers is not as reliable as one using 37 or 43 markers. The higher the resolution of test the easier it is to compare results.

Current academic studies cite an average rate of mutation for the average marker of 1 in 500 cross-generation transmissions of DNA, though some test firms believe the rate is closer to 1 in 350. Accurate figures per marker are some way from being finalized, though it is accepted that some markers do mutate faster than others. I have up-to-date details on these issues on the website that accompanies this book.

The standard statistical tool used to quantify the degree of difference and similarity between two DNA results is highly complex and is known as the Most Recent Common Ancestor calculation, or MRCA. Fortunately there is a simpler, qualitative rule of thumb that non-statisticians can use for everyday analysis and I'll introduce that in Chapter 12 when we look at interpreting DNA results.

In the next chapter we will use the basic ideas that we have covered so far, look first at the different types of test scenario and then at the kind of genealogical research problems that Y-chromosome testing has been able to solve.

More info at <www.DNAandFamilyHistory.com>

- Capabilities of the four online results' databases

- Issues when inputting result data into online results' databases

- Marker and mutation rate information

Chapter summary

• Use the four free-to-use online results' databases to learn more about your DNA signature

• Revisit the online databases annually to see if you can reassess your haplotype

• In a medium resolution test, an exact match on all markers by two men sharing the same surname generally implies that they share a common male ancestor within a genealogically relevant time-frame

• Remain sceptical of family tree linkages suggested solely by the results from low-resolution tests

CHAPTER 5

Y-Chromosome Study Scenarios

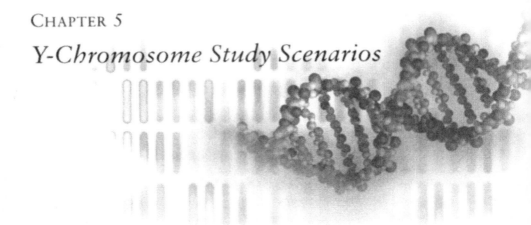

Three types of Y-chromosome study

There are three family history research scenarios that lend themselves to investigation with a genealogy-driven Y-chromosome study. These are:

1. Ancestry problem-solving studies.
2. Surname studies.
3. Studies of a clan, a caste, a tribe, a race or other specific group.

Problem-solving studies are by far the easiest type of DNA study to run. The goal is usually highly targeted and they leave open the option to convert to a full-blown surname study at a later date.

Surname-based DNA studies aim to develop a matrix of Y-chromosome haplotypes for the bearers of a single surname or of a group of related surnames. This is the most popular type of DNA study. The handbook section (Part III of this book) is devoted to explaining in detail how to set up and run problem-solving and surname DNA studies.

The third type of study seeks to aggregate individual DNA results together for reasons other than to create a purely surname-based Y-chromosome signature matrix. Some of these hybrid projects – for example studies looking at clans and castes – work by pooling together the results from many individual surname studies. Others – for example tribe, race and location studies – try to identify a specific attribute that will define these wider groups irrespective of the surnames of the men being DNA tested.

Ancestry problem-solving studies

Many of the most successful Y-chromosome studies to date were set up to prove or disprove a proposition about a group of ancestors or to confirm a long-standing genealogical thesis that could not be proved through documentation. This type of study has produced some extremely useful results, especially for participants who have unusual surnames. In some cases a particular family tree linkage question that has been frustrating or dividing researchers for a long period of time has been resolved with a degree of closure and satisfaction all round.

The most simple of the problem-solving scenarios is to show that the descendants of two or more men bearing the same surname share a single common ancestor. A Y-chromosome test result demonstrating that they share an identical DNA signature provides the best proof of common ancestry – short of finding the absent documentation – that one could ask for.

Some of the most elegant studies have focused on a specific type of genealogical 'bottleneck' in an attempt to jump across a gap in a family tree caused by missing documentary records. Several studies have, for example, tested the modern-day descendants of a group of immigrants (who have the same or similar surnames) whose ancestors arrived in their new host country at around the same time. Several of these studies have then expanded into a second phase of testing to identify descendants who have the same DNA signature and who live in the home country that the original emigrants set out from.

To launch a DNA study it may be easier to target a few researchers whom you know have specific documentary problems that they want to resolve or who are eager to prove a suspected link between two separate parts of a family tree. If those participants find that they have the same DNA signature then at a minimum their personal research will now be directed towards the search for final documentary proof. Conversely, should the expected match of the DNA results reveal that different genetic material has been handed down to the two descendants then this discovery should stimulate a re-examination of the family tree that had previously linked them together. You can safely promote the problem-solving approach to potential participants at the outset of a DNA study knowing that each man's DNA result automatically contributes to the overall matrix of DNA signatures for the surname as a whole.

Case study: proving a single common ancestor

Richard **Steadham**, whose family stems from Alabama, is one of the founders of the Timen Stiddem Society, the family association for the descendants of a seventeenth-century Swedish barber surgeon who settled in New Sweden (now Wilmington), Delaware. His primary concern was to identify which modern descendants bearing the surname Stidham, Steadham or fifteen other surnames are genetically linked to Timen Stiddem's five sons.

Initial results reveal that nine male family members tested who have documented histories linking them to three of Timen's five sons have near-identical DNA signatures while a further four men documentarily linked to one Samuel Stidham, who has not been documentarily linked to Timen Stiddem, do not share this haplotype.

Surname studies

The surname project is proving to be the most popular scenario among DNA study organizers and in many ways it is the easiest to organize, explain and promote.

Y-chromosome surname studies can be limited to just a single surname or widened to include obvious variant spellings. Every multiple surname study aims to identify which of many potentially linked or variant surnames will be confirmed as linked together within both family trees and 'genetic families'. They have even been set up to cover a single Soundex category. (Soundex is a method of phonetic coding used to classify similar sounding surnames.)

A surname study has as its goal the definition of a matrix of DNA signature haplotypes. Such a study can be run for a surname where no genealogical research has previously been done, in which case the DNA results stand on their own as a guide for future documentary research. Every DNA study organizer needs to remember that what the genealogists really want to know is exactly how they are related one to another, not just whether they are related. In a surname study, DNA testing is simply a means towards the end of creating and verifying documented family trees.

One advantage of expanding a single-focus problem-solving study into a second-phase whole-surname study is that if the DNA results' matrix is already in place for families from a particular geographical area they can be used to link not only the DNA results of other people living in that area but also descendants wherever they may be in the world.

One common study arrangement is to create an initial DNA haplotype list for British-origin families and then to map this against the DNA results for emigrants' descendants, most of whom will have emigrated to the former dominions in North America, Australasia and southern and eastern Africa. However, this arrangement can be run equally well in the reverse direction by initially building up the DNA matrix of North American families and then seeking British-origin name-bearers for additional testing. In some cases the latter route may be more successful for DNA study organizers to take, as American genealogists are more receptive to the idea of DNA testing and appear to be readier to invest in a test at an early stage.

No major surname is likely in the near future to declare that its DNA study is closed, or that its DNA signature framework is so well defined and its documented family trees so clearly corroborated that further DNA testing is not required. On the other hand, the number of participants that are needed to take a DNA test in order to declare that a useful haplotype matrix exists may not be as large as one might at first think. The number of results you need to define the matrix for your group really depends on how rare your surname is and whether you have expanded it beyond a single surname to include variant spellings or not.

As an example, while the extended **Pomeroy** DNA study currently has 70 participants, only one of the 19 men tested within the past year and a half has returned a low-resolution DNA result that had not already been found among the previous test participants. Put another way, by sampling around 5–10% of adult males in Britain with this surname we've been able to create a matrix of DNA results that has quickly become a reliable tool to cluster future participants.

The largest surname-framework projects now under way show how powerful this approach can be. As of early 2004, at least eight surname-based projects – for the surnames Bassett, Bolling, Donald, Graves, Hill, Rose, Walker and Wells – had already tested more than 100 participants each. All of these have developed useful clusters of DNA signatures for the purposes of matching future participants.

As the DNA signature matrix for your surname takes shape it will gradually become clear whether your surname is more likely to have

---

Case study: defining a surname DNA matrix

The **Mumma** surname project Y-chromosome tested 40 men called Mumma, Momma, and other variant spellings who live in the USA, Canada, Germany and Estonia. These men link back to 14 different documented ancestors. Of the 40 DNA results, 18 were identical when compared at 25 markers whilst a further 19 differed from this modal result at just a single mutation at a single marker. These results suggest very strongly that there is a single common ancestor for all the members of extended Mumma family however the surname is spelled.

Conversely, the Pomeroy surname study has analyzed 70 men with five different surnames. Seven low-resolution haplotypes were found more than once while the modal result was found only 12 times. This suggests that modern name-bearers have multiple ancestors. Interestingly, the DNA haplotypes did not at all divide along lines based on the spelling of the surname.

---

had a single male ancestor whom you all hold in common or to have been founded by multiple male ancestors. While many potential participants are likely to find the opportunity to prove that they belong to a surname that has a single common ancestor quite appealing, you can promote the multiple-ancestor scenario to them on the grounds that they could find some new relatives whom they could expect to fit into their family tree. Conversely, you can promote the single-ancestor scenario on the basis that the DNA test result will confirm their link to the main family.

Other types of aggregated test programme

DNA study organizers have come up with a number of variations on the theme of a surname DNA signature study, the most successful of which has been to extend it in scope to aggregate the results from a large number of surnames.

Two studies tracking the Irish and Scottish clans are producing interesting results by pulling together results from dozens of surnames. In both countries large groups of surnames are historically associated with specific clans and the English pattern of surname adoption and transmission was disrupted by a combination of warfare, legal restraint and the hangover of traditional naming habits.

One output from these clan studies will be to reveal whether the different surnames that are associated with each clan turn out to share the same genetic material. It will also be interesting to see which of the Irish clans have genetically non-Irish descendants and which Celtic reputations turn out to have been built upon typically Scandinavian DNA profiles.

Apart from the clan studies, there are two other types of wider study that attempt to aggregate results: broadly speaking these are either location-specific or caste-specific studies. My view is that these studies, particularly the location studies, are unlikely to produce conclusive results.

Case study: aggregated result programmes

Results from dozens of surname projects are being aggregated through an online clearing house into a global Scottish clans' study. By pooling hundreds and potentially thousands of individual DNA results, an overall DNA signature matrix will be created that will be able to link any person of Scottish descent who subsequently takes a Y-chromosome test, regardless of his surname.

It will also lead the way in the use of DNA results to identify those haplotypes that have a recognizable geographical affinity with the successive populations that have moved to and settled in Scotland.

Examples of location-type studies include projects testing the descendants of people who lived in remote European locations several centuries ago. This type of project is designed as a quasi-anthropological study, and it will need a larger number of results to create a meaningful DNA signature matrix for a location than is required for a single surname. In order to demonstrate uniqueness you'd also want to compare the DNA results with those of the descendants of families from neighbouring villages.

This comparison problem is less severe with studies seeking to identify the Y-chromosome characteristics of a particular race, ethnic group or caste. This kind of study mimics certain academic studies that have already been successful in showing, for example, a set of low-resolution haplotypes associated with the Cohanim or Jewish priestly caste.

Examples of private projects of this type include one seeking to identify the background of the Melungeons, a distinctive group of settlers of the eastern seaboard of the USA, and another the ancestry of rare castes of the Indian subcontinent.

This type of non-surname study is hard to market even within the genealogy community. In fact, I think in future we'll move to thinking of location and geographical origin as an attribute that belongs to an ancestor, participant or a surname, but not as the organizing principle of a DNA study.

How to set goals to build up your study project

I'd emphasize that it is crucial to define your goals and to think how you can expand your project through several phases before you start promoting your DNA study. If your project is too vague you'll find it difficult to persuade people to pay the money to take a test. If on the other hand you are too ambitious, you will not deliver successful results in a quick enough time-frame to persuade more men to join your study.

It is much easier to market your programme to potential participants if they can see clearly why it is being run and how their individual test result, which they have to pay for, will contribute to the bigger picture. Also, because your programme is likely to be running for several years, it helps to set several milestones within the study so that you can demonstrate as you go along that progress is being made. Keeping up the momentum is very important. Each milestone allows you to update your results, publicize your programme afresh, communicate again with the existing participants, and generate some excitement about the project among the wider surname group. Guidance on all of these promotional and presentational issues is laid out in Chapter 13.

In the next three chapters we'll look in detail at examples of each of the three types of Y-chromosome study that I've outlined above.

More info at <www.DNAandFamilyHistory.com>

• Exemplary Y-chromosome studies

• Online database of Y-chromosome surname studies

Chapter summary

- Define clear goals for your DNA study before you start to promote it

- Plan the study in phases in order to help you to maintain interest among the participants

- A good way to start a DNA study is to focus on solving a specific ancestor question and then to expand it into a full surname DNA study

- As your surname's results' matrix starts to be defined and its single- or multiple-ancestor origin becomes clear, promote this to potential participants

CHAPTER 6

Single-Ancestor Studies

Single or multiple ancestors?

Many genealogical DNA studies have more than one goal in mind and it's not uncommon for these to change as the project grows. However, there is one question that every study seeks to answer, either directly or indirectly: does my surname stem from a single ancestor or from multiple ancestors who started using my surname at different times?

Genetic tests are a highly useful tool to identify linkages among a distinct group of people, especially when you expect that their DNA results should be identical or highly similar. This scenario is most likely to be true:

- When the men are members of a single documented family where the DNA tests are being taken to confirm those links.
- Where the men share an unusual surname.

The great simplicity of the single-ancestor study is that it is extremely easy to promote to potential participants whilst leaving the way open for it to be turned later into a full surname study.

Single-ancestor and problem-solving study goals

I've highlighted in this chapter examples of single-ancestor and problem-solving studies as well as the single-ancestor aspects of whole-surname studies. The basic types of goal that they seek are:

1. To provide proof, or disproof, of a connection with a known historical person, e.g. a clan chieftain, a first immigrant, or a famous fellow name-bearer.
2. To establish whether two documented but unconnected family trees whose descendants share the same surname are in fact related through a common genetic ancestor within a time-frame that accounts for their shared surname.
3. To unravel family trees that have become jumbled up with ancestors bearing different variant surnames.
4. To test a number of people with the same – or close variant – surname in both the countries of immigration and emigration in order to match together a family in the New World with its ancestral family in Europe.
5. To identify the surname of the suspected father of an illegitimate male child in a documented family tree by DNA testing both sets of descendants.

As you can see, these goals cover a broad spectrum of research cases. However, in all instances the key is to establish the DNA signature of the target line both accurately and quickly.

As with any documented family tree, one usually needs to DNA test several members in order to establish the authentic DNA signature of the common ancestor at the head of it beyond reasonable doubt. Of course, this number increases if all the results don't match as expected, as may often happen in very large family trees. Once identified, this haplotype then acts as the reference result for everyone else within the rest of the DNA study. This basic method applies also when seeking to prove that two distinct family trees are one, or to unravel various lines in them that appear to be mixed up.

A programme to link surname-holders in the original emigrant country with their descendants in the New World is really a surname study in all but name. However, in the case of rare surnames, where the thesis of the project is that everyone bearing it is related through a single common ancestor, the project is much more logically explained to participants as a single-ancestor or problem-solving study.

The fifth case I listed above involving a suspected illegitimacy within a documented family tree is much more complicated and in practice very difficult to resolve satisfactorily. Not only must the DNA signature of the descendants of the main surname-bearing family be clearly identified, but you will also need to have identified more than one male descendant of the suspected father and to have secured their agreement to take a DNA test. What's easier to do is to test several members of the original family tree so that you are able to show that the descendant from the suspected illegitimate relationship does not have the same DNA signature.

Let's look at some actual examples to see how these elements are applied in practice and how they blend together within individual DNA studies.

Case studies

There's nothing quite like the allure of a well-known historical person or colourful documented family history to pull in participants to a DNA study.

A good example is the **Campbell** DNA project where the declared purpose is to identify the DNA signatures for as many as possible different modern Campbell lines, including a reference set belonging to the present-day Duke of Argyll. It's worth noting in passing that genetic testing is an especially powerful tool for surnames where traditional documentation may not exist or where groups have switched surnames at a specific point in history. This was often the case in families that stood to inherit a title. One of the targets of the **Boyd** DNA surname programme, for example, is to establish a genetic connection with the descendants of James Boyd who changed his surname to Hay when he became the 15th Earl of Errol in 1758.

The attraction of the historical connection was also a factor in the **Pomeroy** study when it was first launched in 2000. One of the carrots I held out to potential participants was that I would be able to check their DNA result against that of a known descendant of the Norman noble family who first held the name. I admit that at the time I thought of this more as a marketing feature than as a real aim of the project, and although I did not expect to find any connections among our participants to my surprise when the results came in no fewer than 5 of 70 were similar enough at a low resolution that I could tell them that it is conceivable that they could in fact share a common ancestry with the noble family within the past 1,000 years.

Even more surprisingly, one of the five had, prior to the testing, shared a story with me about how their link originated, a tale that he said had run in their family for years. I confess that I now pay a bit more attention to the oral histories I hear even though I doubt sufficient documentary evidence will ever be found to prove or disprove these particular possibilities (though, of course, testing additional markers might provide compelling genetic evidence).

The original immigrant ancestor scenario has a similar emotional pull for potential study participants, particularly in the USA. In many cases the arrival of the immigrant marked an effective genetic bottleneck in the history of that surname, that is to say the modern-day descendants in that country of settlement stem from that single immigrant and form a single coherent family tree. Tracking the genetic descendants of the original settler is a powerful testing proposition in countries as large as the USA and Australia, but it can apply to smaller nations also.

There is an original immigrant element within the global Pomeroy study. Over the past 450 years several hundred family members have emigrated from England to the five most important former British colonies. The earliest known of these was one Eltweed Pomeroy who left Dorset for Massachusetts most likely in 1632. Today, thousands of Pomeroys in the USA know themselves to be Eltweed's lineal descendants and their extended family is the largest in our worldwide family today.

Within the global Pomeroy surname DNA project I set up a sub-project to track Eltweed's descendants. One participant opted for a DNA test in order to check a family story that one of his ancestors was illegitimate, a story that the DNA evidence backs up. Several other descendants have identical low-resolution haplotypes but in a few cases show one or two single-step mutational differences at higher resolutions. As the extended family is well documented it may prove possible, if sufficient participants come forward for testing, to be able to pinpoint where in the tree the particular mutations took place. The **Mumma** project cited in the previous chapter has already been able to isolate a specific mutation for one of its main descendant lines.

A variation on this kind of study seeks to link two documented but unconnected family trees. In some cases this also entails showing that different surname spellings belong to the same genetic family.

One example is the **Allred** DNA project where the family is attempting to identify the DNA signatures of four men who arrived in North Carolina in the 1750s. Among Allred researchers these four men are referred to as brothers, but no documents or any hint of their parentage have ever been found. In this case, if identical DNA signatures are found from each of the four lines this would confirm the possibility that the four men genuinely were brothers.

The very strong modal result found in the **Mumma** project is clear evidence that the extended Mumma family, including those bearing surnames that are variant spellings, share a single common ancestor. A geographical breakdown of the results shows that, regardless of the spelling of their surname, men tested from the USA, Canada and Germany share a common ancestor while those from the Estonian branch do not.

DNA testing is an extremely fertile way of generating hints for future documentary research. In some cases the transmission of a surname can become extremely entangled within a family tree so that it's simply not clear who was married to whom. In other cases several documented families that were living within the same small locality might have had male members who were all given the same forename, each of whom heads up a line of descendants that can't be reliably linked back. The challenge here is to work out which of the permutations of marriages and births are consistent with the DNA results of their descendants. In this type of case it can actually be helpful that there are some mutations present in the results as these can act as guidelines during the family tree reconstruction process.

The **Stiddem** DNA study already mentioned on page 64 is an excellent example of what can be achieved. One of the goals was to identify the DNA signature of the Scottish immigrant John Steedman, whose descendants married into the Stidham family in eighteenth-century South Carolina and took on a variety of surnames with slightly different spellings. These DNA signatures found were used to match modern Stidham and Steedman surname-holders within the main family tree to confirm and also to correct their documented ancestry. Of the two documented descendants of John Steedman who have taken part in the study, one man who today spells his surname Steadman turned out to be a direct descendant of Timen Stiddem not John Steedman as he had always thought.

Identifying an illegitimate ancestor

A number of DNA studies have an illegitimacy issue at the heart of one their family trees that they hope that Y-chromosome testing may be able to resolve. The crucial factor in this type of study is that a descendant of the man thought responsible for fathering a boy in the main surname's line should be identifiable and willing to take a DNA test to demonstrate the fact of the connection.

The Mumma project had a sub-project looking at a case like this. In one documented tree a descendant with the surname Moomaw was recorded with a DNA signature markedly different from the Mumma modal haplotype. A review of the documentary research suggested that back in the 1850s his direct male ancestor had been born to a woman who had been born a Mumaw. This ancestor was traced as a child to an entry in a census where he was bearing the surname Webb. As the family of the putative father had already been researched it only required a low-resolution test taken by a descendant of that Mr Webb to demonstrate that his DNA signature was identical to that of his descendant today bearing the name Mumaw.

While such closure is still unusual, there are many variations on this particular project theme. The origins of an eighteenth-century American immigrant Captain Daniel **Little** are the subject of a specific project run by researcher David Roper. From documentation it is believed that Captain Little hailed from Europe and that it was likely that he had changed his name on arrival in the USA. After some research two options have been put forward; firstly, that he was originally the son of Johann Heinrich Klein of Zweibrucken in Germany, and secondly, that he was the son of Sebastian Cline, born in the Palatinate on the border of France and Germany.

Although the answer has not yet been found, indications of a link with either a Klein or a Cline descendant may become visible when a number of DNA signatures for these two surnames have been recorded. Given that both of these names are relatively common in the USA, finding a connection by this broadly based method is perhaps a bit of a long shot. However, as the total number of Y-chromosome test results rises, this broad approach should start to come into its own even with common surnames.

Remarkably, the Mumma project reports that one man with a different suname but with a known illegitimate male ancestor was able to piece together a plausible documentary trail to demonstrate a link with a specific Mumma ancestor based on the exact higher resolution match of their unusual haplotypes.

A note on illegitimacy, infidelity and surname changes

All DNA programmes will, at some point, have to grapple with what geneticists in euphemistic fashion describe as a 'non-paternity event'. In the context of genetics this describes the introduction of new male genetic material into an ancestral line by someone other than the expected male parent – something which should perhaps more properly be called an 'extra-paternity event'.

The impact of this in a surname study could hardly be greater. Instead of a matching or closely mutated DNA signature, the male descendants are likely to have a completely different haplotype altogether. When charting the DNA signatures in a surname study this new interloping material often sticks out like a sore thumb. It looks so different from the modal haplotype that it cannot possibly be the product of genetic mutation within a meaningful time-frame.

DNA results do not lie, and if the difference between two DNA signatures cannot be accounted for by mutational change then clearly something else has caused it. However, our ancestors chose to change their surnames much more frequently than we might perhaps imagine. In these cases the documented family trees would appear intact while the genetic material will be different; in other cases the family tree is changed whilst the DNA signature remains the same.

There were many reasons why a surname might be changed, or a new family line created, or the DNA signature changed in an existing family, not all of which are shocking. Among the scenarios are:

1. An illegitimate male birth to an unwed mother, where the male child takes his mother's surname (and may be brought up as her parents' late child).
2. A birth within a marriage where the male child is fathered by a man other than the wife's husband.
3. A male child from a previous marriage of the wife of a surname holder, the step relationship being revised by the formal adoption by the child of the stepfather's surname.
4. A surname change demanded of the male descendants of a married female member of the family in order to ensure that family property is inherited within the surname
5. The use of an alias which in time becomes the main name

associated with this particular line, and hence its surname.

6. Personal choice, for example where a man marries an heiress wife and adopts her surname.

7. As a result of the misspelling of the surname to a point where it no longer looks like a variant spelling of the original name.

None of these changes has been researched to the point where we have quantified estimates about how often these events are likely to crop up in the average family tree. Keep it in mind, however, that even a low average rate of non-paternity events (for example, 1 in 100) will eventually disrupt the orderliness of the 'genetic' tree.

A key point to consider in relation to illegitimacy and DNA results is that the interloping DNA may not in fact look very different from the main family's DNA signature. It may well be that in some isolated parts of Britain many men in a given geographical area may have very similar haplotypes. This concentration of genetic material would have happened long before surnames become fixed.

Genealogists should also realize that a null match between two DNA signatures does not necessarily negate a documented tree that links those two descendants together. With only two DNA results, particularly if they belong to descendants that fall on opposite sides of a large family tree, it is difficult to know where the slippage in DNA transmission occurred and why. The only way to find out is to test more male descendants within the tree in order to narrow down the options.

From the last few pages it will be clear that a genealogist can have as much, if not more, fun with a set of unmatched DNA results as one can with a set of results that match together as expected. And as so often happens with documentary research, the answers that arise tend to generate new questions.

More info at <www.DNAandFamilyHistory.com>

- Single-ancestor case studies

- Dislocations between 'genetic trees' and 'family trees'

- Data on non-paternity rates

Chapter summary

• A well-defined single-ancestor or problem-solving project is an excellent way to launch a DNA study

• The main organizing principle is to try to demonstrate genetic connections where there is no documentary evidence to prove a family tree link

• Specific ancestor studies can be given an additional dimension by comparing the DNA results of surname-bearers in different parts of the world, and especially in making linkages between host countries in Europe and emigrant families in North America

• A great deal more work needs to be done to quantify how and why genetic material and surnames are not always transmitted in tandem

CHAPTER 7

Surname Studies

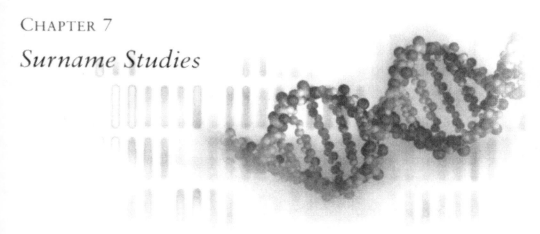

Analyzing Y-chromosome results by surname

At its simplest level, a surname DNA study consists of aggregating as many DNA signature results from men with the same surname as possible and analyzing the patterns of genetic connection that emerge. This usually produces immediately useful results because the genetic similarities they reveal highlight previously unsuspected links between members of different family trees and suggest further lines of documentary research.

While a worldwide study of a unique surname is the easiest type of surname study to run, there are a number of different elements and stages that it can incorporate in order to develop its potential. These include:

1. Tracking a single surname in a specific geographical area rather than globally.
2. Including variant, related surnames as part of a multi-surname study.
3. Including foreign versions of the principal surname being studied.
4. Focusing on countries of emigration and immigration.
5. Widening the study to include all surnames categorized within the main surname's Soundex category.
6. Linking any of items (1) to (5) to a background set of documented family trees.

You'll notice that many of these surname studies chime well with the different types of studies I categorized in the previous chapter as specific ancestor or problem-solving studies. In fact, the line between them and a surname study is not precise. As a DNA study collects more results it will inevitably start to look like a fully-fledged surname study even if that was not the original intention.

What's important from the point of view of the study organizer is that a single- or multi-surname study requires a more nuanced approach to the analysis of its results. The most crucial of these is analyzing whether the surname has a single- or a multiple-ancestor origin.

Single ancestor or multiple ancestors?

In a single-ancestor study, the main goal is to identify the unique haplotype of that ancestor. A surname study is more complex: a surname, or group of surnames, may link back to a single ancestor, or it may not. As we saw in Chapter 3, it is logical to expect that most surnames would show multiple-ancestor origins, but we've already seen examples in the previous chapter of surnames that clearly are of single-ancestor origin.

One of the goals of a surname study is to understand, as early in the study as possible, whether the modern bearers of a surname do link back to a single ancestor or whether many men adopted the surname over the years. A strong concentration of results around a single haplotype – a strongly modal pattern – argues for a single ancestor, whereas a diverse set of results where no single haplotype appears to be dominant suggests a multiple-ancestor origin.

While the boundary between these two patterns, modal and multiple, is undefined, in practice most sets of surname results can be categorized as one or the other. The key to resolving the question is to make sure that you test enough participants, using medium- or high-resolution tests, so that you are able to make a secure judgement, taking into account the frequency of the surname in the general population.

As a personal rule of thumb, in a surname group with, say, 50 men randomly tested, if more than 50% have the same haplotype or a similar one it would be reasonable to believe that the surname stemmed from a single ancestor. Similarly, if the most popular haplotype was found in only 10 men, which is just 20% of the total, one would take up the hypothesis that the surname probably has multiple ancestors, especially so if there were several haplotypes all equally common. Note that this rule of thumb only applies to men randomly selected; biasing the sample by, for example, testing several men from the same family tree or geographical region might produce a different result entirely.

While every case needs to be corroborated by documentary evidence, the main challenge is to interpret the origins of those surnames that fall between those 20% and 50% benchmarks. As the study of Y-chromosome results and surnames is still so young, reliable statistics do not yet exist, and there is certainly going to be a great deal of debate and dispute before this issue is resolved. At the moment the best advice any study organizer whose surname results fall into this grey zone can be given is that you should persuade more name-bearers to take a Y-chromosome test.

There are several components in this modal versus multiple debate. Key among them are the resolution of the haplotypes being compared, the number of participants in the study and the type of surname being studied. Common sense would suggest that rare, low-frequency surnames will turn out to have a single ancestor whilst common, high-frequency surnames will have a great number of originating families and hence multiple ancestors. However, we've already seen from the Sykes study that this is not always clear-cut. Professor Bryan Sykes, who defined the original single-ancestor origin thesis with the results of the Sykes surname study, firmly promotes the modal model of Y-chromosome distribution within English surnames. In the first chapter of his latest book, *Adam's Curse*, entitled 'The Original Mr Sykes', he returns to this issue to ask:

> Had I been incredibly lucky with the name Sykes? I don't think so. Over the past two years I have replicated this study with dozens of names. Not all of them show as tight an association between surname and Y-chromosome as Sykes, but most do...

His conclusion, that 'the majority of surnames, in England anyway, are very clearly linked to one or a very few Y-chromosomes', is surely going to be revisited and challenged during the next few years. It will be interesting to see if this 'dominant single-ancestor thesis' is corroborated by the hundreds of surname studies now under way worldwide.

Tracking a single surname in a specific geographical area

Very common surnames are extremely hard to study just by documentary means even in a single country let alone on a global basis. The number of surname-bearers is simply too great to be linked together with any success. However, Y-chromosome testing does offer a way forward if high-resolution tests are used. These, however, have only recently been offered on the market, so it is too early yet to see the useful results for the most common surnames.

The other way to tackle this problem is to restrict a surname project to one geographical area. Rather than tackle a worldwide project for a surname that accounts for more than 1% of all men of British-origin, one **Smith** project is tracking only the Smiths of the southern states in the USA while a separate partner project covers the north-eastern states. The linkage process is rapidly grouping their participants together and stimulating the documenting of their family trees.

Similarly the **Turner** project is focusing on linkages that exist among the numerous Turner lines that have migrated from the mid-Atlantic region of the eastern USA, particularly families that moved west and south from Virginia and Maryland, rather than attempt a national US or global project.

Multi–surname studies, including variants

Many DNA studies are built around a group of surnames that have equal status with each other even though some are more common than others. The **Pomeroy** study includes several linked surnames – namely Pomroy, Pomery and Pummery – that are found in Britain and that have links to the same geographical areas as Pomeroy. It also includes two other surnames that are today found only outside of the UK – Pommeroy in Australia and Pumroy in the USA – both of which are understood to stem from the same root.

Interestingly, the genetic families grouped by the Y-chromosome results include men with different surnames. This confirms something that many genealogists know only too well, that the spelling of a surname is an unreliable guide when reconstructing family trees.

In fact, the spelling issue is more complicated than that. Documentary research has already shown that certain spellings are common in particular English counties; for example, Pomery is the most common spelling in Cornwall whilst Pomeroy is prevalent in Devon. Of course, the surnames in many families have altered over the centuries, a process that has often introduced the dominant modern spelling of Pomeroy into families that trace back to a Pomery ancestor. Initially the results looked very confusing as men with different surnames were found to have the same low-resolution DNA signature. But when looked at again as the first high resolution test results came in, the Y-chromosome results suggest that the modern-day bearers of the surname Pomery link back to a very small number of documented families. Furthermore, all the families with a Cornish-origin man surnamed Pomery at their head conform to the pattern of a single ancestor, i.e. they look from the genetic evidence as though they are all related. These results suggest that most men surnamed Pomery have a single common ancestor even though men surnamed Pomeroy do not.

Multi–surname studies, including foreign equivalents

This type of study is increasingly popular among US-based projects where many project organizers are looking at surnames introduced into North America by multiple immigrants and where 'surname naturalization' has occurred as foreign names have been anglicized upon entry.

A good example is the **Cooper** study which includes the variants Couper and Cowper (from Britain), Coupard and Cuopard (from France), and Cuper, Keefer, Kieffer, Kifer, Kueffer, Kupfer and Kupper (from Germany).

The linked surnames can either be brought into the study on the basis of a semantic linkage – for example the German and French equivalents of the English profession of 'cooper' – or because they sound like the English surname Cooper and might plausibly have been adopted by immigrants to the USA.

Emigrant/immigrant studies

This type of surname project, which is designed as a targeted international surname study in order to track both sides of the emigrant/immigrant divide, is also popular among US-led research groups.

The **Long/Lang** project is built upon a lot of documentary research into families called Long, mainly from Wiltshire in England, who emigrated to the USA in the early days of colonization and have now spread throughout it. They were joined later by French Longs (originally De Long) and German Longs (originally Lang). This study started by refining the various surname groups in the USA by using genetic evidence to create new linkages between them. In a second phase it will expand to identify the specific areas of Europe where these immigrant groups originated.

Soundex studies

Some research groups have found it difficult to identify the boundaries between the different potential variant surnames and as a result some Y-chromosome studies have opted to test anyone whose name falls within an entire Soundex category.

The **Brazil** project includes anyone with a name that falls within Soundex code B-624, i.e. any surname beginning with the letter 'B' and sounding like 'dazzle'. It reports that identical DNA signatures have been found in men with the surnames Braswell, Brazell, Brazil and Bracewell.

Surname studies backed by documented family trees

The gold standard among Y-chromosome studies is the marriage of a long-term worldwide documentary project with a new Y-chromosome test component. This kind of surname DNA study can be extremely powerful in its reach both to confirm long-accepted family trees and to knock holes in them.

The Pennington Family History Association has data on more than 16,500 **Pennington** and variant surname households worldwide. The 56 Y-chromosome results they have collected to date reveal firstly that several distinct families have adopted the surname Pennington over the years – i.e. it is a multi-ancestor surname – and secondly that all of the currently documented families that have held the surname for many centuries probably link back to just one or two villages in Lancashire in north-east England.

There are many examples of this kind of DNA/documentary combined study, including the **Rice** study, led by the Edmund Rice (1638) Association, and the **Payne** study, which discovered that well-documented English Payne families from Suffolk, Huntingdonshire and Jersey in the Channel Islands are members of a single family, not unrelated families as was previously thought.

Inspiring surname programmes

Y-chromosome surname studies are in their early days still, but there are at least eight surname projects, all investigating high-frequency surnames, that already have over 100 participants. These are studying the surnames Bassett, Bolling, Donald, Graves, Hill, Rose, Walker and Wells.

The **Graves** study is well worth reviewing in detail on organizer Ken Graves' website at **<www.gravesfa.org/dna.html>**, in particular the charts that outline the possible ancestral connections that the results have prompted and the section on the success stories in his programme where he describes how previously unconnected family trees have been linked together.

The **Walker** study at **<http://freepages.genealogy.rootsweb.com/~fabercove>** has found no fewer than 27 groups in its 195 DNA test participants. Information about the ancestral lines is grouped by DNA signature on the website, making this a useful tool for Walker researchers everywhere to monitor progress of the study.

The laurels are reserved for the US-based **Wells** study at **<www.wells.org>** organized by Orin Wells, a long-time member of the Guild of One-Name Studies, which now has more than 300 participants. The project was built around 24 documented families and has expanded from its original focus on US families to include British families too. This shows the classic profile of a growing surname DNA project where the results suggest new lines of documentary research and identify new potential test participants in a repeated result/research process.

Links to these three and other studies can be found on the website accompanying this book.

Bridging the documentary gap

Such large-scale global surname projects probably look daunting to anyone approaching Y-chromosome testing for the first time, but it's worthwhile repeating that many of the most useful benefits produced by genetic testing are best accessed by small projects and readily occur at an early stage of the project's development.

One example from the Pomeroy study illustrates this point, and it's a tale that will be familiar to many genealogists everywhere: the breaking down of a persistent research 'brick wall'.

My example concerns an American researcher who had researched his Irish parentage back to County Cork, in southern Ireland, in the 1750s. After two decades of fruitless searching for the final piece of documentary evidence to link his ancestor to the Pomeroy family that was known to be living in the area at the time it looked as though all he had to 'prove' his link with them was his own family's oral history and coincidence. That was until we realized that our DNA project had generated a haplotype for the Cork family based upon the low-resolution tests of two men who can reliably trace their link to that family tree. After a nail-biting time waiting for the DNA result to come back from the lab, our American researcher was relieved and delighted to find that his result was an exact match on all the markers they'd been tested on in common. Even though the shared haplotype is not a rare one and the matching was only able to be made at low resolution, the fact that the family history predicted the DNA result that was found adds weight to the argument that it was not coincidence.

Challenging existing documentary work

DNA results generally throw up so much new data and stimulate so much fresh thinking that it is as well to note that they regularly require study organizers to re-examine family trees which have long been held by their researchers to be definitively established.

Every genealogist who has been researching for a while will be familiar with the process whereby family trees shift imperceptibly over time from a state of speculative linkage to established truth. The more they are shared with other researchers the more often they are accepted as fact. Sometimes DNA testing puts an incontrovertible question mark over unwittingly falsified family trees and the subsequent re-examination of them illustrates how powerful DNA testing is as a tool to correct and restore them to an accurate state. I have no reason to believe that our documentary research is atypical, but I have seen this scenario worked out several times already in the Pomeroy surname study and I expect to see it again before too long.

In one instance, several participants who had been DNA tested appeared to link back to families with origins in the English county of Wiltshire. It became clear, however, that two participants who were documented as part of the same family tree had different DNA signatures whilst a third who was not connected had a DNA signature in common with one of this pair. The most likely explanation is that two family lines have been crossed over at some point and confused by a researcher. After looking back through the trees – and buying some marriage certificates – the solution became apparent: two men called Henry Pomeroy had had their marriages and descendants wrongly attributed to each other, a piece of guesswork uncovered by the DNA results and subsequently resolved by documentary corroboration.

The most compelling and successful genealogical studies integrate their DNA test results with their ongoing reconstruction of family trees. Each set of new Y-chromosome results stimulates an iterative process where the family trees, built using documentary evidence, have their accuracy cross-referenced by the results from DNA testing. As the study progresses the DNA results identify potential mistaken linkages in the documented trees and suggest areas where missing documentary links could be sought. They not only clarify lines of research and assumptions made during the tree-building process, they also confirm current assumptions and flag problematic ones.

More info at <www.DNAandFamilyHistory.com>

• Surname case studies

• The 'Modal versus Multiple Origins' debate

• Analysis of single-ancestor and multiple-ancestor surnames

Chapter summary

• Surname studies come in different flavours and their Y-chromosome results can quickly generate powerful benefits for genealogists focused on surname reconstructions

• DNA testing reveals whether surnames have a single ancestor or multiple ancestors

• With around 1,500 surname DNA projects declared in mid-2004, the largest study under way has over 300 participants

• DNA testing is best seen as a way of checking and refreshing documentary research, not as a substitute for it

CHAPTER 8

Clan and Caste Studies

Extended DNA projects

Some DNA study organizers have set up large-scale projects that resemble academic phylogeographical studies. These are investigating the genetic profiles of clans, castes and even the inhabitants of particular geographical regions. They are, however, all quite different in nature.

The big clan projects – tracking members of Irish and Scottish clans – are pooling the results of a large number of participants who are already taking part in surname studies as well as finding participants directly by advertising a large number of associated surnames under the banner of a single clan. These clan projects look to me to have a good chance of producing very interesting results indeed. In most cases potential participants already recognize the clan affiliation of their surname and the scope of the project is easy to understand. There is no reason why over time they will not expand to include any man with a surname of Irish or Scottish origin. Two key outputs of the studies will be to reveal whether there is much genetic difference between the Scottish and Irish clans, and to discover which clans in the two countries are relatively more closely linked to each other across the Irish Sea.

The situation with caste studies, and with others that seek to identify either ethnic origins or migrant groups, appears to me to be far less certain. All of the DNA testing companies are organizing their marketing, study group assistance and results' databases at the level of the surname. While many surnames have a distinct regional distribution, many features that add up to an ethnic origin belong primarily to individuals and not to surnames or regions. Some studies that are targeting particular groups, such as Indian castes, are, however, using the surname as the trigger for inclusion in the study to some effect.

Another group of DNA studies are targeting the descendants of families that used to live in a specific geographical area. These are much less likely to produce clear-cut results. This type of test is difficult to market and most of the online databases of Y-chromosome results currently available are weak in terms of the location data they attach to individual test results. Regional data on the frequency of different haplotypes are more likely to emerge as a result of improvements to the DNA results' databases rather than from specialist geographical studies like these.

The Irish clans study

The Irish clans study has grown out of an academic testing programme using the genetics lab at Trinity College Dublin, sponsored by a member of the **Guinness** family. It is now expanding from its initial phase, which looked at a tight group of potentially related surnames and clans, to a follow-up phase that will expand the testing to include a wider range of Irish surnames.

The original purpose of this study was to discover if there are genetic similarities, as expected, between members of the Magennis and Guinness families, but it also included bearers of the surnames McCreesh, Neeson and McNiece that are linked to these two families as well as associated surnames such as Donleavy and McCartan that lived in areas neighbouring the Magennis country (between Lough Neagh and the city of Newry in the northern part of Ireland) almost a millennium ago.

The early results have identified several distinct 'genetic families' within the Magennis surname, including one that is recognized as defining the lineage of the clan chief. Fortunately this haplotype is quite rare so it stood out clearly during the analysis, at two millennia the oldest genetic/lineage connection yet discovered. Interestingly a similar DNA signature was found in the neighbouring McCartans which may confirm that the two clans shared their ancestry some 1,400 years ago.

Surnames tests are under way on a number of individual Irish surnames including **McTiernan, O'Rourke** and **O'Donaghue**. The latter project, for example, has 35 results to date and is tracking the descendants who originated from each of eight ancient tribal areas associated with the O'Donoghues in Ireland. Considerable migration from these areas took place over the centuries and family groups took root in many other counties in Ireland, which many present-day members now mistakenly recognize as their county of origin.

The **O'Shea** project covers this surname and other related kinship groups including the surnames O'Connell and O'Falvey. The name O'Shea traces back to an eighth-century ancestor. About 400 years later a branch migrated away from its original settlement area in county Kerry in south-western Ireland and moved east into Tipperary and Kilkenny, so the project is testing members of the three branches to see if they share a common ancestor.

Scottish clans studies

The Scottish Clans project, which is run through the Scots-DNA list on Rootsweb by John Hanson, takes a broad aggregation approach. It is pooling all the Y-chromosome results for any surname with a Scottish connection whether incorporated into a clan study or not. As with the Irish clans, many study organizers have recognized the potential benefits of DNA testing for Scottish families and by spring 2004 more than 30 surname projects had been registered with the prefix 'Mc-' or 'Mac-', and many more will surely follow.

DNA testing offers an exciting opportunity to provide a degree of definition and resolution to a series of historical debates about clans and lineages. Scottish clans originated in the kin groups of the Picts and the native Scots, later becoming overlaid with feudal obligations and subsequently associated with the clan chief or laird who was in turn known by the name of his chief estate. While clans were formed from septs or kinship groups – each of which could have formed its own surname – and which are assumed to be descendants of common progenitors, the leadership of the clan and its inner core of clansmen was not necessarily passed on by heredity. The exact features of clan and kinship are often debated and Professor David Hey notes in *The Oxford Companion to Local and Family History* that a clan was 'a unit that contained families of different lineages; common descent is often assumed but cannot be proved…chiefs, like the barons of England, were indifferent to whether or not the people over whom they ruled shared the name and blood.' Long-term conflicts between clans sometimes led to entire clan names being erased from public life across several generations, so a single lineage might change its surname more than once in a way that simply didn't happen south of the border.

There are many large-scale Scottish clan projects under way. The Clan **MacGregor** study follows the main surname back to the original four-teenth-century Gregor, though some genealogies go back further to King Alpin whose descendants are said to have become the founders of the MacKinnon, MacQuarrie, MacAulay and MacAlpine names. The sur-name was proscribed from 1603 to 1775, when it was illegal to use the name in official documents. Many families did not retake the clan name when it became legal again to do so, and as a result the clan now recog-nizes more septs and aliases than most other clans, surnames that are as different as Bennett, Black, Dowie, Greer, Gregory, Gruer, McAdam, McLaughlin, Gilmore, Nevins, Orr, Shankland and Stirling.

The DNA study has over 100 members to date. The results show that the common ancestor for the modal haplotype lived some time before the fourteenth-century clan foundation date, raising the possibility that some of the different lines were at that time headed by members of a group of kinsmen rather than by a single individual. Early clansmen might well be related by blood, but they might also be related through community, that is by living together and adopting the surname of the principal leader of the community. Those not related by blood are often referred to as 'part-takers'. It had always been assumed that because the clan was so persecuted only 'true' MacGregors would bear the name. Nevertheless, it is clear from the DNA that Clan Gregor does have its ori-gin in one individual but also that it contains 'part-takers' sharing a com-mon Celtic culture – probably of Irish origin – as well as 'part-takers' whose paternal ancestry is Icelandic or Viking.

The complexity of the family interconnections and the richness of the history of the **Donald** clan are another good example of the genealogical potential of major clan DNA studies. The clan history links the families of MacDonald, MacAlistair and MacDougall, all three of which claim to be descendants of a king called Somerled. By legend he was a descendant of the Irish kings, his ancestors having taken over lands in western Scotland in the sixth century. Despite having a Norse-origin name, Somerled freed much of the western isles from Norse rule before his death in 1164.

The Clan **Donald** DNA project, registered with a testing company called Family Tree DNA, includes no fewer than 59 surnames and had tested 147 members as of May 2004. The surnames included range from Beaton to Houston, from Balloch to Murdoch, and from MacBride to MacSporran. Indeed the study will include anyone who believes that their ancestor originated in the historic Clan Donald land, i.e. the western highlands of Scotland including Kintyre and Ross, the Inner and Outer Hebrides, or Ulster, Leinster or County Derry in Ireland.

While the public Clan Donald project has yet to publish its results, Professor Bryan Sykes has also conducted Y-chromosome tests on members of the MacDonald clan – including five clan chieftains – with the goal, as he relates in *Adam's Curse*, of identifying 'the Y-chromosome of Somerled himself'. The haplotype he identifies is rare in Scotland but common in Scandinavia, leading one to suspect that the ancestor was of Norse or Viking ancestry. Given an estimated number of perhaps 2 million male MacDonalds worldwide, Professor Sykes speculates that as many as 400,000 of them could have 'Somerled's Y-chromosome' some 36 generations later.

Caste studies

Some Y-chromosome projects have been put together that target a relatively small group of people who are linked together by a common birthright that suggests a degree of common ancestry. They hope that their results will show a strongly modal distribution which will then identify the shared heritage of their caste.

This type of quasi-anthropological project needs many more participants than a standard surname study if it is to produce a meaningful result simply because its participants' results are not merely being compared against each other but against the general population. At present it is very difficult to collect this kind of background DNA material, so in effect half of the project may not be attempted.

One of the earliest projects is the **Chitpavans** project led by Dr Jay Dixit. The members of the Chitpavan – or Kokanastha Brahmin – community number slightly over half a million in India, a fraction of 1% of that country's total population. Less than 300 years ago the community was strictly isolated in a small area around Chiplun in the area of

Konkan in the state of Maharashtra. Even today, it is estimated that fewer than 15,000 Chitpavans live outside India. While no surname is an exclusive indicator of community membership, the Chitpavans themselves recognize at least a dozen surnames as typical as well as maintaining strong oral traditions of community membership.

Another subcontinent project is dubbed the **Emperor of India** Surname Project. This seeks to 'bring together the last surviving descendants of His Majesty Abu Zafar Suraj ul Deen Bahadur Shah Ghazi II', the last emperor of India who was deposed by the British in 1857. It is using surnames Babur, Mughul and Timur as the primary indicator of potential membership.

Ethnic group studies

Caste studies have the potential to produce useful results as caste members are generally linked over an extended period by a common birthright. Ethnic group projects are, however, at another remove from the commonality implied in a shared surname. Each particular project needs to be judged on its own merits, of course, but it's hard to imagine how multisurname projects like this will produce useful results, except perhaps to show which low-resolution haplotypes are common in a particular region.

An example is the **PA Deutsch Ethnic Group** DNA project, which focuses on the descendants of Pennsylvania Germans – also known as the Pennsylvania Dutch – where the goal is to determine whether the group has more genetic material that can be classified as 'Asian' than average, thus in turn confirming stories of its origins to the Huns and Mongol hordes that invaded Europe during the previous millennium.

Another established community project is that investigating the heritage of the **Melungeons**, a mixed-race group from the eastern seaboard of the USA, which has a strong set of oral traditions but no clear idea of where its people originated. This has published some anonymized data but as yet no report.

Regional and locality studies

Some studies have been launched by individuals using Y-chromosome tests to discover ancestral links between men bearing different surnames.

While it might be useful to aggregate results based upon the assumed geographical origin of specific surnames, it is difficult to build a stand-alone DNA project this way.

Two such studies focus on areas of Britain that used a patronymic naming system – where the father's given name was passed on as a sur-name – as recently as 200 years ago. The **Welsh Patronymics** project is basically an aggregation project like the Scots Clans project except that it takes any recognizably Welsh-origin surname.

A second project is reviewing the DNA – both Y-chromosome and mtDNA – of men with links to the **Shetland Islands** that lie to the far north of Scotland. However, although an exact Y-chromosome match on 37 markers would strongly indicate that both participants share a common male ancestor within the past few hundred years, it gets one no nearer to working out when and how they are related. While it would be interesting to see whether men with Shetland connections who have Norse-origin names – such as Williamson, Anderson or Thomason – have different types of Y-chromosome signatures compared to men with Scottish surnames – such as Leask, Bain or Muir – the surname link is simply a more modern feature.

Other projects attempt to define themselves by race, e.g. investigating the descendants of colonial and Creole inhabitants of Puerto Rico based upon matching DNA and surnames, or an isolated geographical area, e.g. men whose ancestors lived in small sub-Carpathian (eastern European) or Italian villages. Again, it's hard to see how these projects will gather sufficient results to declare a thesis.

'Deep ancestry' studies

Commercial testing companies recognize the demand from many of their clients who want more detail about their ancient ancestral origins, or 'deep ancestry' as described in Chapter 2. Some companies now offer Y-chromosome tests that suggest that they can identify an aspect of one's ethnic or tribal background. At different times test companies have sold tests to link their clients to the Jewish priestly caste, the Vikings, and to high-profile individuals such as Genghis Khan. However, no incontrovertible genetic test for any type of ethnic or tribal membership currently exists. Furthermore, the results may not turn out as one might expect.

Roughly 30% of black American men have a European haplotype, according to the testing firm African Ancestry. All this kind of test does is to check particular haplotypes or marker values and look for the best match in the results databases of modern populations. Put another way, finding that your haplotype is common in Sweden does not necessarily make your ancestor a Viking. Furthermore, virtually all of the reference data were derived from low-resolution tests. In the current climate where results from high-resolution tests are calling into question some of the assumptions built up in the past from low-resolution results, some caution may be in order.

Space precludes me from looking at these in detail in this book, but I've highlighted them on the accompanying website along with a list of academic studies that this kind of test references. I'll return to these tests when considering how you should choose a testing company in Chapter 10.

A note on mtDNA and autosomal DNA tests

The mitochondrial and autosomal DNA tests mentioned briefly in Chapter 1 find their proper place here in a book aimed at genealogists for the simple reason that this kind of 'ancestor heritage' test is designed to reveal data about our deep ancestral origins, rather than about connections that occurred in our recent past, i.e. within the past millennium. While the results can be seen as challenging to those who have received them, they simply do not help the genealogist's main endeavour of recreating family trees built around a shared surname.

Mitochondrial DNA is passed down the maternal line and reveals data about that single passage of inheritance in a mirror image of male line through the Y-chromosome test. Autosomal tests use the DNA from both parents that gets mixed up as it passes from one generation to the next – differently from the non-recombining Y-chromosome and mtDNA that is used respectively in lineage and deep-ancestry studies. An autosomal test quantifies in percentage terms the breakdown of your total DNA from all of your many thousands of ancestors. To a classical genealogist bent on reconstructing family trees this is about as useful as knowing one's star sign. The accompanying website holds further details of these types of test should you wish to know more.

Having looked in some detail at the different factors one needs to bear in mind when analyzing DNA results, and the type of DNA studies being run by genealogists, we can now turn to look in detail at the practical considerations of running a DNA study.

More info at <www.DNAandFamilyHistory.com>

• Progress reports on aggregated group studies

• Details about 'deep ancestry' tests using mtDNA and autosomal DNA

• List of academic papers referenced

Chapter summary

• Studies that pool together the DNA results of many surnames are more complex to organize than single-surname studies

• The most exciting multi-surname results are coming from the Scottish and Irish clans projects

• Tests of some very specific groups, such as small-population Indian castes, may yield useful haplotype results, but location-based studies are unlikely to do so

• mtDNA and autosomal tests produce little benefit for the classical genealogist

PART III

THE DNA TEST ORGANIZER'S HANDBOOK

9 *The DNA Test Checklist*

10 *How to Select a DNA Testing Company*

11 *How to Launch and Market Your DNA Study*

12 *How to Analyze Your DNA Results*

13 *How to Present and Publish Your Study*

14 *The Future of Genetic Genealogy*

CHAPTER 9

The DNA Test Checklist

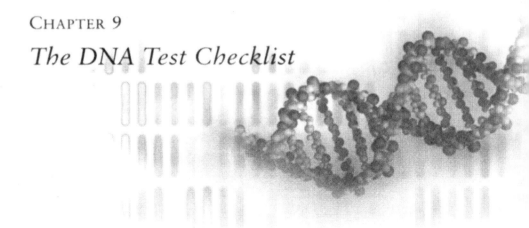

Six steps

If you have an internet connection and a credit card, taking a DNA test is really very straightforward. All the test companies are able to register your order online and to take payment by credit card. They will also answer your questions by email or phone before you place an order and can issue you an invoice by post if you wish.

There is no reason why someone in Europe should not use an American testing company, or vice versa, as the swab kit used in the test is small and light enough to be sent through the regular mail. Anyone can

---

Figure 5 The six steps in the DNA test process

1. Define your test goals.

2. Select a test company and order a test.

3. Return the supplied swabs with your DNA sample.

4. Receive your results and report from the company.

5. Compare with the online results' databases.

6. Review against your genealogical research.

physically take their own DNA sample completely painlessly in just a few seconds: the process is really that simple.

Define your test goals

For most genealogists, the primary goal of taking a DNA test will be to identify new members of your family tree or to confirm existing ones. As family trees are traditionally built around the father-to-son transmission of a surname, you will need to select a Y-chromosome DNA test as this tracks your genetic inheritance down the paternal line.

If you want to set up a DNA study to group together the results of everyone with your surname you'll do best to register with one of the testing companies that have facilities to help you organize and promote your study. With some testing firms you also have the option to take a follow-up Y-chromosome test – known as a haplogroup test – that will give you an indication of your deep ancestral background, or to receive information about your likely haplogroup based on your haplotype result. Full details about how to select a company and to set up and run a DNA study are set out in Chapters 10 to 13.

Although it is highly unlikely to be of any direct use to any genealogical research that you are undertaking, if you want to find out about your deep ancestral background as handed down to you through your direct maternal line, all the major testing companies offer a basic mitochondrial DNA (mtDNA) test that will do this for you.

Select your test company and choose a test

If a surname DNA study already exists for your surname it makes sense to use the same testing firm that others have already selected simply because it will be slightly easier for you all to compare your results.

The current crop of testing companies offer tests that measure anywhere between 9 and 43 markers. As a rule of thumb, a 24-marker test is the minimum that you should select. A low-resolution test with 9, 10 or 12 markers is of little use to genealogists, except in the single case that you want to prove that your DNA signature is the same as the already known DNA signature of someone else in your family tree and where additionally you know that this specific DNA signature is a rare one in Europe (assuming you have a name of European origin).

You can compare value for money between the different testing companies by calculating the cost per marker of the tests they offer. As we'll review in the next chapter, the current cost of Y-chromosome tests ranges from a high of around US$32 per marker tested to a low of under US$5.

Return the supplied swabs with your DNA sample

Once you've signed up for a DNA test your chosen company will mail you a small swab kit consisting of several medical-type cotton sticks and a consent form.

The cotton sticks need to be scraped gently around the inside of your cheek for at least 30 seconds each in order to pick up some of the microscopically small cells lining the skin. There is no discomfort involved. It is best to use the swab sticks at different times, and you must in any case ensure that you've not cleaned your teeth or drunk a milky drink for several hours beforehand. First thing in the morning is a good time to take your sample.

The consent form varies from one testing company to another. At a minimum it will ask for your consent to test your DNA, but it might also ask your consent to store your DNA sample, to retain your DNA results for future study, or to enter your DNA results on a database held by the testing company. (Note that when you receive your results back several of the testing companies may then recommend that you add your results to a free-to-use public database of DNA signatures affiliated with them.)

Receive your results and report

The DNA testing company will post and/or email a written report of your results to you. This will detail the exact numerical allele values per marker that they have obtained from the DNA sample that you provided on the swabs, plus any comments they want to make about the data and the results.

If you have email they may also email you notification that other men in their internal database of DNA results are an exact match with your

DNA signature. While this is useful to give you an indication of the types of surnames that match your DNA signature, it is extremely unlikely that you are related to any of these men within any genealogical time-frame if you do not share the same surname or a recognizable variant of it.

Compare with the online results' databases

You can learn more about the distribution of your DNA signature by inputting and/or checking your DNA results with the four public resource results databases as discussed in Chapter 4. You can search these databases for free to see who else has registered their DNA results and who shares your DNA signature. These websites also publish information about the markers they use and the distribution of allele values per marker. This can be useful if you want to start making a more detailed analysis of your results, for example to see whether your results for each marker are rare or common.

Use the academic YHRD database at <www.yhrd.org> to view the geographical distribution of your Y-chromosome haplotype in the samples submitted by academic researchers in Europe, North America and Asia. While only 8 of the 9 markers used in the YHRD database are generally used by the genealogically oriented DNA testing companies, the results will still give you a good indication of the frequency of your low-resolution DNA signature together with the broad pattern of its distribution across these three continents.

Review your genealogical research

If you're lucky you may find someone else who shares your surname or has one like it, and whose DNA results are identical or close enough to suggest that you could be related within a genealogical time-frame. This should stimulate you both to review your family trees to try and find the male ancestor through which you could be mutually linked. Don't raise your expectations though as many people can only trace their family trees back, say, no more than two centuries while the shared DNA signature may signal a familial connection between the two of you that may have occurred several centuries earlier. Even if you cannot see an immediate link in your family trees, the DNA result indicates that you may find a documentable family tree connection if you search for it.

Bear in mind that you can only really be confident that two men with matching DNA signatures actually belong to the same family tree if you have found an exact match using a 24-marker or a higher-resolution test. If you exactly match on a 10 or 12 marker test you can nonetheless start your family tree research with the hypothesis that you are directly related within the same tree, but you should not assume that this is the case. It is quite possible that, if you have a high-frequency surname and your DNA signature is relatively common, you may belong to two separate trees but still share the same surname and low-resolution haplotype. Conversely, if your surname is relatively rare and your DNA signature is also rare, you have a much greater degree of assurance that you are both related and are part of the same, single family tree.

As your documentary research develops and you find more people to whom you might be related, you can in some cases ask your DNA testing company to reprocess your original DNA sample and expand the number of markers tested. With a higher-resolution result you stand a better chance of confidently rejecting any near low-resolution match or of confirming the possibility that you are all members of the same family tree.

How to organize a DNA surname study

The real benefit of a DNA test is to compare your results with other men with whom you might be related, i.e. as part of a surname-based DNA study.

There are no qualifying criteria for DNA study organizers and if you look at the profiles of people running them around the world they are a very diverse group of men and women. Technically, you do need access to the internet and to be comfortable with using email, but other than that you just need to have a degree of intellectual curiosity, the willingness to deal with strangers, and patience. You'll find that interpreting the results is not as complex as you might fear and that the testing companies are ready to assist you if you need help. Beware though that, like all genealogical activities, running a DNA study can suck up your time like it's running out of style!

I'm often asked, 'if Y-chromosome tests can only be taken by men, what is the female genealogist to do?' The answer should be clear at this point, and many Y-chromosome projects are run (and funded) by female members of a family history group.

Organizing a surname study can be as simple or as complicated as you wish to make it. At a minimum you can just announce a surname study through your chosen testing company and wait to see who enrols. However, there are many active steps that you can take beyond this to promote your study worldwide and turn it into an absorbing historical analysis. I explain the ways in which you can do this in Chapters 11 and 13, but next let's set some guidelines to help you choose a testing company.

More info at <www.DNAandFamilyHistory.com>

• Updates to the DNA checklist

Chapter summary

• Choose a medium-resolution 24-marker Y-chromosome test at the minimum

• You can safely choose a test company based outside your own country

• Check your results on the public online databases to see who else matches your result and where your haplotype has been recorded

• Organizing a surname DNA study is challenging, rewarding and straightforward

CHAPTER 10

How to Select a DNA Testing Company

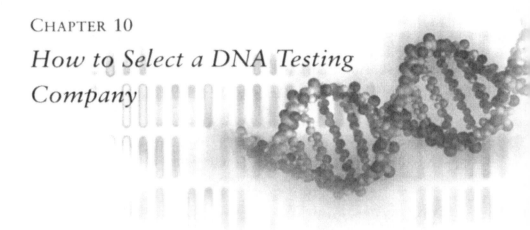

Organizing a DNA surname study

Organizing a DNA surname study need not be very taxing. Several test companies offer considerable online resources to assist a would-be study organizer and the best are able to help you at every step of the way so that you can organize your project efficiently and easily share the results within your group.

There are three key factors to be taken into consideration when you choose which testing company to team up with. These cover:

- The degree of online assistance they can give you to organize the project, including the scope of the databases of results they use.
- The cost per marker of the tests offered to your research group and the total number of markers that are measured in their tests.
- Their ability to store your DNA sample so that you have the option to have it retested for additional markers later on.

There are no fewer than 11 companies offering DNA tests that are aimed both specifically at genealogists and at the wider audience of people who are interested in finding out more about their ancestral roots.

Only six offer the standard Y-chromosome test that is required in a DNA surname study. Of these six only three – in alphabetical order, DNA Heritage, Family Tree DNA and Relative Genetics – offer the range of services required by a genealogist planning to undertake a surname study.

DNA Heritage

DNA Heritage is a British company based in Weymouth, Dorset. It sent out its first test kit in October 2003. Its founder, Alastair Greenshields, earlier launched YBase, the first free-to-use public database of Y-chromosome haplotype results that appeared on the web. His goal is quality service: 'I'm far more interested in having a few well-informed, satisfied customers,' he comments, 'than in testing high numbers.' Prices are identical for group members and individuals. The company's website offers a great deal of help to surname study organizers and has a small number of surname study projects registered.

In spring 2004, the 43-marker test DNA Heritage was offering was the highest-resolution Y-chromosome test available anywhere in the world. Unlike other testing firms, DNA Heritage allows you to select any combination of markers so that you can compare results from different test companies more easily, and to exclude markers that do not appear to vary much.

It does not offer to retain DNA samples for retesting as it does not expect this to be necessary. Initially using an academic lab in a university, DNA Heritage now uses the Sorenson Genomics lab in the USA to process its DNA samples.

Website: <www.dnaheritage.com>
Email: info@dnaheritage.com
Phone: +44 (0)1305 834936

Family Tree DNA

Family Tree DNA (FTDNA) is an American company based in Houston, Texas. It sent out its first test kit in April 2000 and is currently the global market leader in terms of the number of tests sold worldwide. Its founder, Bennett Greenspan, has a reputation within the DNA testing community of being readily available to help his customers and the company has the capability to preserve your DNA samples.

Family Tree DNA uses the lab of Dr Michael Hammer of the University of Arizona (who is also chair of the Y-Chromosome Consortium) to run its tests. A leading statistician in the genetics field, Dr Bruce Walsh, sits on the company's advisory board.

It offers a range of Y-chromosome tests at 12, 25 and 37 markers. The company's website offers a great deal of help to surname study organizers, including a web page to promote it, and has over 1,000 surname study projects registered as under way.

Family Tree DNA's internal database combines the results of its clients with academic data sourced from the University of Arizona. It currently holds some 22,000 Y-chromosome records. The company funds the public free-to-use YSearch results' database to which its clients can upload their results with a single click.

Family Tree DNA is the only test company that offers to make follow-up tests, price US$100, to confirm your haplogroup according the Y-Chromosome Consortium (YCC) definition. Even better, its internal database of results is often able to predict the haplogroup of your haplotype. This identification is included as part of your results' package free of charge.

Family Tree DNA's low-resolution 12-marker Y-chromosome test is also sold through the website of Ancestry.com.

Website: <www.familytreedna.com>
Email: info@familytreedna.com
Phone: +1 (713) 868-1438

Relative Genetics

Relative Genetics is an American company based in Salt Lake City, Utah. It sent out its first test kit in August 2001. The company is funded by James Sorenson who also set up Sorenson Genomics, the in-house lab that Relative Genetics now uses to handle its DNA tests. The company is the only lab in the world that is ISO 17025-certified by the US National Forensic Science Technology Center for all of its genetic genealogy testing services.

Relative Genetics offers a high-resolution 37-marker Y-chromosome test (measuring 43 alleles) and a medium-resolution 24-marker test (measuring 25 alleles). The firm offers an additional enhanced-level service that includes detailed analysis of results by an in-house team of genealogists. Clients can elect to store their DNA sample for future retesting for up to 25 years.

The company's website offers many resources to help group study managers. The database of results that the company uses, through the

Sorenson Molecular Genealogy Foundation, contains more than 20,000 haplotypes. The company is currently supporting an estimated 250 registered surname study projects.

The Relative Genetics 24-marker genealogy test is also sold through GeneTree, which is a subsidiary of Sorenson Genomics that specializes in paternity testing.

Website: <www.relativegenetics.com>
Email: info@relativegenetics.com
Phone: +1 (801) 461-9760

Other companies offering Y-chromosome tests

The other three testing companies that offer a standard Y-chromosome test are either targeting groups with specific origins or only offer low-resolution tests.

Oxford Ancestors at <www.oxfordancestors.com> is a British company based in Oxford that was spun out from Oxford University by its founder, Professor Bryan Sykes, author of the only academic paper on the use of DNA testing and surname groups. When it sent out its first test kit in February 2000 it was visibly the first company active in the new genetic genealogy market in Britain and it still performs its lab work under its own roof. However, it only offers a low-resolution 10-marker Y-chromosome test at £180 (US$325) and does not offer website support for surname study organizers. It is actively promoting a test that claims to tell you whether you are a descendant of Genghis Khan.

African Ancestry at <www.africanancestry.com> is a specialist testing firm that targets its tests towards the African-American segment of the US market. It offers a 9-marker low resolution Y-chromosome test for US$349 (£193) which is analyzed by Sorenson Genomics. Its head, Dr Rick Kittles, of Howard University, Washington, has put together a proprietary database of results from African DNA samples called the African Lineage Database with over 10,000 Y-chromosome results that it claims that can pinpoint the geographical location in Africa of a paternal ancestor. It does not handle group studies.

GenByGen at <www.genbygen.de> is a German DNA testing firm based in Göttingen. It offers a 9-marker low-resolution Y-chromosome test for Euro 199 (£133). Its website is in German only.

Companies offering mtDNA and autosomal DNA tests

It can be noted that mitochondrial DNA tests are offered by a wider range of companies in addition to the six firms profiled already. These include GeoGene at <www.geogene.com>, Roots For Real at <www.rootsforreal.com> and Trace Genetics at <www.tracegenetics.com>. The latter specifically targets the Native American segment of the market.

Autosomal DNA tests are marketed by two firms, Ancestry by DNA at <www.ancestrybydna.com> and DNAPrint Genomics at <www.dnaprint.com>. This kind of test aims to quantify your total ancestry by stating a simple percentage figure for the geographical components in your total genetic make-up. Neither of these types of test is any use for genealogists constructing traditional family trees through paternal lines.

Criteria for selecting a Y-chromosome testing company

With three mainstream companies offering broadly similar and competitively priced Y-chromosome tests, how is the would-be DNA study organizer to choose between them? Based on my own experience, I outline here the key criteria I personally think that a surname study organizer should use when selecting a testing company to partner with over the long term (see Figure 6).

Five of the six companies I've already highlighted have plans to improve their services, and in some cases to offer new prices, during the autumn of 2004. To review these improvements and to see the latest comparative evaluation of the testing companies, please check this book's website.

Number of markers measured

The greater the number of markers that you test, the higher the resolution your DNA signature haplotype will be and the more confident can be your judgements when comparing DNA test results.

A few years ago a 25-marker test was the gold standard for the few family historians embarking upon a DNA surname study. Now it is

Figure 6 Five key criteria when choosing a DNA testing company

1. Number of markers measured.

2. Test price offered to surname group participants.

3. Size of the firm's internal database of results.

4. Level of assistance for surname study organizers.

5. Option to retest further markers using the original DNA sample.

becoming generally accepted within the genealogical community that no Y-chromosome test with fewer than 24 markers is worth promoting as part of a wide surname-type study. The low-resolution tests simply cannot be relied upon to distinguish between individual families with sufficient clarity, especially when the number of men being tested within a surname group increases above a few dozen. At low resolutions the most common DNA signatures appear to lump together many families whose documented histories suggest that they are unlikely to be directly related as part of a single family within a genealogical time-frame. Rare DNA signatures, of course, often stand out even in a low-resolution test, so low-resolution tests are by no means without use. However, it's only with tests of 24 markers and upwards that one can be relatively sure that the individual 'genetic families' will differentiate themselves clearly one from another to the point where you can use the results with confidence to direct your documentary research.

The number of markers required to generate genealogically useful results for a surname study is still a contentious point and one that will come into clearer focus during the next few years as the number of surname studies reporting their results increases. What is already clear is that there is an absolute limit on the number of markers that need be tested. If one tests too many markers then an exact match would be a rare event; if one tests too few markers then many separate genetic families will appear to merge into a single one. Put another way, as the number of markers increases the cost of the higher-resolution tests continues to grow but without providing you the client with any extra useful information. With the highest-resolution tests now testing either 37 or 43 markers, this is probably about as good as it will get with the current type of Y-chromosome test. I'll return to this issue in Chapter 12 when discussing how to analyze Y-chromosome results.

Test price offered to surname group participants

Most test companies offer lower prices to men taking their DNA test as a member of a declared surname study compared to men signing up individually for a one-off test. However, each testing company sets its own qualification criteria for group membership and the level of price discount it is prepared to offer. During the past few years prices have generally been quoted only by the testing company itself directly, but the increasing competition between them is leading to a degree of price variation at each level of test resolution depending on where you purchase the test.

In short, the price picture may well become less clear over the next few years and the best advice I can give is that the would-be surname study organizer should shop around on the internet. It is especially worthwhile to watch the exchange rate of UK sterling against the US dollar as the prices offered by non-UK testing firms – which quote their prices in dollars – can cut the price dramatically for British genealogists, and potentially vice versa. Again, check this book's website for an up-to-date list of prices.

Figure 7 uses the prices companies were quoting for non-group tests in July 2004 in order to illustrate the relative value-for-money they offer.

Size of the firm's database of results

Clearly it's in your interest as a study organizer to work with the testing company that has the largest possible reference set of Y-chromosome results to compare with your own. The companies themselves see this as crucial and are working hard to secure access to the data sets created and used in the academic world in a bid to make their offer to their clients more appealing.

They already build up their advertised databases from many different sources. The primary source is always the results of their own clients.

Figure 7 Y-chromosome tests/value for money

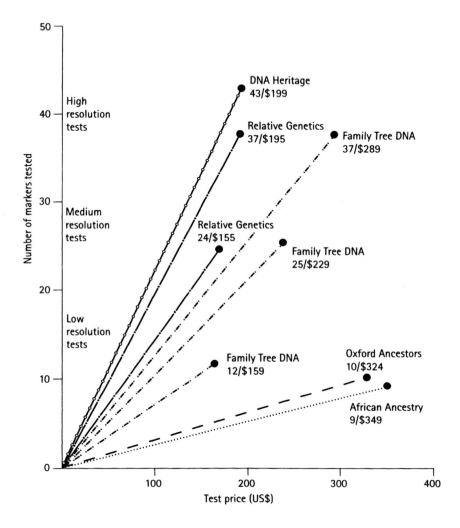

Notes
1. *These are prices for an individual test. Group test prices, where offered, are cheaper.*
2. *Some companies test markers that return more than a single allele value.*
3. *Conversion rate £1.00 = $1.80.*

Secondary sources are the reference sets that their academic partners have researched, plus any other academic data sets they can access. Finally, some companies also reference results that can be seen in the public free-to-use databases and from other public sources such as published academic papers, and potentially even websites where study organizers report their results.

In general, in order to maximize the benefits to you when comparing results, you'd be best advised to choose the company that advertises the largest number of results that it uses for comparison purposes. Bear in mind though that almost all the academic reference sets are low-resolution haplotypes, while the genealogical community is primarily interested in medium- or high-resolution haplotypes.

In addition, as we saw in Chapter 4, anyone on the web can access the four public free-to-view databases that contain Y-chromosome test results. This is worthwhile to do even if, as three do, the public databases have an affiliation with a specific testing company.

Group study assistance

As a surname group study organizer, the more assistance you can get from a testing company the better. Ideally you're looking for a mechanism to announce that your surname test study is under way, to canvass support for it (marketing and promotion), to sign up men who want to take the test as part of your programme (sales), to collate your results (service and analysis support) and to communicate with your group members as you develop the study (aftercare).

DNA Heritage and Family Tree DNA have designed their websites to offer group study organizers a private area to review details about your project's members and their DNA results, and to perform the activities listed above. From within the organizers' area you can post information to your group's participants and review the results as a group. The same companies have put a lot of background resources into the public sections of their websites, and both they and Oxford Ancestors run online discussion forums where their clients can ask questions.

Family Tree DNA publishes a monthly email newsletter while DNA Heritage publishes one from time to time. These two have good online resources that explain background information about the markers used in their tests and some of the nuances of interpreting DNA signature results. Relative Genetics links to similarly useful information on the Sorenson Molecular Genealogy Foundation's website.

Other selection criteria

Figure 8 outlines five further criteria that form part of what I see as being the 'Top 10 criteria' when you're selecting a testing company partner. While the first 3 criteria of the 10 are quantitative, the remaining 7 are well worth considering even though they are largely qualitative and often subjective.

All test companies can handle online test ordering by credit card. You may meet problems if you want to pay for a test from an American-based firm with a British cheque or Switch card, but your best bet is to talk to the testing companies directly by phone or email and ask their advice. They can also advise on how to apply for a DNA test by post.

The most organized test companies – at this time DNA Heritage and Family Tree DNA – offer you the ability to track the progress of your test kit through their system. They allow you to order a test for a new, or an

---

Figure 8 Additional criteria when choosing a DNA test company

1. Online ordering and order integration.

2. Quality of lab procedures and speed of delivery of results.

3. Interpretation and presentation of individual results.

4. Readiness and ability to communicate with study organizers.

5. Overall reputation of the company.

---

upgrade test for an existing, member of your surname group, to track the tests in your group that have already been ordered and are in process, and enable you to present your results in a way that makes it easy for you to communicate them with your group members.

There is no quality-control mark for DNA testing labs that specifically applies to genealogy-type tests, though the Sorenson lab used by Relative Genetics has the ISO 17025 mark. However, all the participating labs will have ensured that their equipment is correctly calibrated to meet international academic and national forensic standards (where appropriate).

The test companies handle the interface between you, as the client, and

Table 4 Comparison of test companies' key capabilities

	DNA Heritage	Family Tree DNA	Relative Genetics	Oxford Ancestors	American Ancestry
Test resolutions offered	High	High/Medium/Low	High/Medium	Low	Low
US$ cost per marker (individual tests)	4.63	7.81–13.25	5.72–6.46	32.40	38.78
Price reductions for groups	Yes	Yes	Yes	Yes	No
Haplogroup tests offered	No	US$100	No	No	No
Free haplogroup estimate	No	Yes	No	No	No
Surname projects registered	Not known	1,000+	250+	Not known	Not known
Internal database of Y-chromosome results	Not known	22,000	20,000	Not known	10,000
Online assistance for surname studies	Yes	Yes	Yes	No	No
DNA storage offered	Yes	Yes	Yes	No	Not known
Online tutorial materials	Yes	Yes	Yes	Yes	No
Newsletter	Yes	Yes	Yes	No	No
Online bulletin board	No	Yes	Not known	Yes	No
Lab used to process results	Sorenson	Univ. of Arizona	Sorenson	OA	Sorenson
Sponsored self-submitted results' database	YBase	YSearch	Sorenson MGD	None	None
Year founded	2003	2000	2001	2000	–

the lab, as the testing body. Clearly there is scope for mistakes when humans are labelling samples and processing results – what has been dubbed by early study organizer Doug Mumma as the potential for 'clerical mutations' – but there's no easy way to judge whether any company performs less than perfectly in this regard.

The turnaround time from the point of ordering the test kit to the delivery of your test results is around 6–10 weeks and can fluctuate depending upon demand regardless of which company you choose.

The clear and accurate presentation of results is crucial for every participant in your study, including the explanation of the statistical calculations and the terms used (such as 'significance levels').

Keep an ear open for news about the reputation of the testing companies and other peoples' comments about how they communicate with their clients. This is tremendously important for you as a study organizer but difficult to measure. The best people to ask are existing study organizers in the companies' online discussion forums, but you can also ask for pointers in the non-affiliated discussion forums listed on this book's website.

Conclusion

Any of the three leading testing companies can be expected to act professionally towards you and to offer a high-resolution Y-chromosome test backed by a range of useful web-based help for you as a study organizer.

You may see advantages in choosing to sign up with the market leader with the most surname projects under way, or the company with the largest database of results, or a company that gives you the option to retain your DNA sample for retesting; it is up to you. As the market develops the companies are aligning themselves closer together in terms of the cost per marker of a test and the choice of any specific company will probably become less critical to the success of your project.

More info at <www.DNAandFamilyHistory.com>

- DNA test price offers from the testing companies

- Latest information about markers, databases and companies' services

- List of online forums where DNA study organizers exchange views

---

Chapter summary

- Review the latest group prices before selecting a test company partner

- Check the test companies' websites for news of changes to the markers they offer and the size of their databases of comparative results

- Check <www.DNAandFamilyHistory.com> regularly for updated information and test price discounts

- Check with other study organizers that the testing company that they use offers the features that you are looking for

CHAPTER 11

How to Launch and Market Your DNA Study

Launching a DNA study

You should expect that running an active DNA surname study will require you to make a commitment to market and promote it in a sustained fashion over several years. As around 25,000 men have signed up for commercial Y-chromosome tests to date, there's no longer the credibility hurdle that existed just a few years ago to jump over. It may still, however, take you some time to persuade the first few participants to sign up for a new surname study, although the amount of DNA-related information now available on the web makes it much easier to explain its benefits to people who are new to the idea.

Many of the international surname DNA studies now under way have arisen from inside long-established documentary genealogy research groups. If you're keen to set up a surname DNA study the first thing you need to check is whether a genealogy research group, which would probably have information on many documented families, already exists, and then secondly, to check whether a DNA study linked to it has already been launched.

There are two types of online searches you can perform to find this out. The first is to check whether the surname is registered with the British-based Guild of One-Name Studies (GOONS) at <www.one-name.org>. If the name is already registered you are in luck. The registrant will likely have a range of contacts with documentary researchers all over the world and should be able to help you reach potential DNA test candidates.

If your surname is not registered you can double-check whether any-one is collating documentary information in a planned way by posting a message on the relevant online surname discussion forums hosted by Rootsweb, Ancestry and GenForum. The other researchers you meet there should be able to tell you if anyone else is already running a region-al or global documentary surname study.

Next you need to check whether anyone else is already coordinating a DNA study for your chosen surname. The two places you need look are:

1. The Genealogy-DNA List Task Force's self-submitted database of surname DNA projects at **<www.dnalist.net>** which includes all projects registered with Family Tree DNA and Relative Genetics.
2. Family Tree DNA's own list of the surname projects registered with their company at **<www.familytreedna.com/surname.asp>**.

If you see that someone is already organizing a DNA test project for your chosen surname, try to avoid setting up a competing one. If the existing study is only aiming at a restricted geographical area you may think it a good idea to set up a complementary project for a different area, but be aware that by splitting the results into two separate projects you will miss some of the key benefits offered by the testing companies that enable you to analyze all the results together. A better idea would be to contact the existing DNA study organizer and propose splitting the organizational and promotional duties within a single study. One big global study shared between two people is much easier to run, and is likely to be more successful, than two smaller regional studies. This also applies if you'd like to run a DNA study for a surname that could be considered to be a variant of a surname already the subject of a DNA study. It is a good idea for all study organizers to appoint a deputy or co-administrator from the outset. This ensures the survival of the project and continuity in the event that the main organizer becomes indisposed.

Recruiting DNA study participants

Email is by far the most efficient and cost-effective way of recruiting par-ticipants to your DNA study. The best way to kick-start it is to compile an email list of known researchers and surname-bearers and to contact them with an explanation of your goals and a clear outline of the expected benefits for everyone researching that surname. You may have to build up this list of 'sales prospects' by trawling through the places where they congregate such as online forums.

If you don't already know the genealogical background of your DNA study's potential participants then seek to collect it right from the outset of the project. You can draft a short questionnaire to send out to each participant that will ensure that you collect the same data from each one. At a minimum this would include questions about their birthplace, the birthplace of their father and paternal grandfather, and notes on any oral history stories relating to the family. It would also be interesting to ask them about their understanding of the origin of the surname simply to view the range of different ideas that they bring forward.

If you're without much data on the family trees associated with your surname you'll basically accept anyone into the study who bears the name. If you already have documented trees to hand you can try and proactively approach descendants who belong in the larger trees. This would get your study off to a good start because the larger trees are the most likely to produce DNA signatures that you can use as well-defined reference points.

Eventually you will reach the point when you will want to test more than one man in each large family tree in order to check that the DNA signature you found with the first participant is valid throughout the entire tree. When doing this try to choose two men who are related as far back in time as is possible. In any case, do not suggest that two close relations – for example, uncle and nephew – take a test together. Even if they are aware of the possibility that their expected DNA match might not be found, the test results of two closely related men will add no value to a whole-surname DNA study.

Marketing your DNA study

Once the study is launched you'll want to use a range of means to publicize and promote it. While you can rely to an extent on passive methods such as word of mouth, in practice if you want your DNA study to grow to a significant size you'll need to employ a range of active marketing measures. Although I'll outline the passive methods first, the active ones will in fact create a far greater number of participants for your study over time.

The principal passive method is simply to list your project in obvious places such as on the list of registered projects maintained on your test company's website. You should also register your DNA test programme with the independent database of DNA surname studies mentioned earlier in this chapter.

You will want to create a presence on the web to promote your project. Family Tree DNA provides basic web space where you can advertise your project as a link from their alphabetical list of studies if you are unable to build your own website. If you can do so, even a one-page website on a free web host – such as Rootsweb's Freepages – can be set up to be indexed by the main search engines so that after a few months it will appear high up in the results of any web search matching your chosen surname and the word 'DNA'. Your web page can be set up to link readers directly to your chosen testing company which will then handle online orders, payments and the dispatching of test kits direct to participants. You can also register with Peter Kramwinckel's web ring of DNA studies which will help to encourage other people to visit your web page, and you can also announce its presence in online forums for your surname or region of study.

You can go one stage further and set up an email newsletter list by using an email package as simple as Outlook Express, but there are also web-based options that will handle the subscribe/unsubscribe process for a simple newsletter.

You might also get some free promotion for your study by offering short articles about your project to genealogical magazines. This is likely to be most effective in the family history magazine most appropriate to the region or county where your surname originated. You may also be able to tie this article in with an announcement on their website and to their own email list.

Documentation

As the study gets under way it is a good idea to set up a set of standard documentation for all participants. This will certainly include background questions about their family tree plus a consent form that gives you permission, as the study organizer, to hold their results, to publish them either separately or as part of family trees, and to upload them onto third-party databases for the purposes of advancing your project.

In the consent form you might also want to include some standard warnings about the undesirability of the DNA testing of siblings together with a line to indemnify you against any consequences stemming from the participants' DNA results and your interpretation of them.

The documentation you send out for marketing purposes should also outline a set of reassurances designed to put potential participants' minds at rest. First among these is that the test results revealed to family historians can tell you nothing about a person's ethnic origin, appearance, intelligence, medical history or susceptibility to any known medical condition. Furthermore, while some testing companies will hold your DNA sample for subsequent reprocessing should you request it, every company allows you to elect to have your sample of DNA destroyed after it has been tested should you so wish.

Working with the results

If your study is registered with any of the three leading testing companies, they will inform other clients in their database whenever one of your study's participants scores a direct match with them. You can use this system to connect the members of your own surname project, if you choose, as well as to identify matched men with other surnames.

When your participants get their results they will want some help to understand their significance and an explanation of how they fit in to the wider surname DNA project. As the study organizer you are the guardian of the integrity of your project. At a minimum this means maintaining a list of the DNA signatures together with an indication of which individuals tested in the programme have which signature, plus background information about their origins and their family trees.

What I've outlined in this chapter is, in fact, just the beginning of a one-name Y-chromosome study. The real goal of DNA testing is to help people to expand their documented family trees, and in a fully-fledged study you will also be tracking and updating the documented family trees of each participant. Of course, you can leave the documenting of the family tree to each participant himself. However, if you do so you'll be missing out on the most engaging aspect of a surname DNA study: linking together the trees of those members who appear to share the same haplotype.

If you are actively researching the family trees of everyone in your surname as well as running the DNA study side of things then your options for analyzing and publishing the combined DNA results and enhanced family tree research are much greater. How to perform the analysis and interpretation of the DNA results is the subject of the next chapter.

More info at <www.DNAandFamilyHistory.com>

- Online promotional resources: forms, forums, databases and software

- Sample documentation, e.g. questionnaire, consent form, emails

- Information on the Data Protection Act for British genealogists

---

Chapter summary

- Define the goals of your study clearly at the outset and promote your study in stages according to the project milestones that you've laid down

- Be prepared to actively market and promote your study both online and offline

- Maintain an email list of your core group of current and potential participants and email them regularly with updates of your study's progress

- Prepare standard research and consent documentation at the start and keep your research paperwork in order including a master list of the results

CHAPTER 12

How to Analyze Your DNA Results

'Well, are we related or not?'

It's a challenge that always arises for a DNA study organizer during the process of analyzing Y-chromosome test results. Two participants, keen to find new research leads and family members, pose the question: 'Well,' they ask, 'are we related or not?'

At this point neither of them wants to hear detailed, nuanced explanations about the percentage probabilities of different scenarios or warnings about the unknown mutation rates of fast- and slow-mutating markers. What they want is a straight answer.

Fortunately, with the arrival of high-resolution tests and the increasing amount of DNA reference data that is available both to the testing companies and to you as a study organizer, it's getting easier to give your participants that straight answer. However, even as you try to satisfy them you need to remember that there is always some room for doubt. If you've been active as a genealogist for even a short while this should be second nature to you, but it's easy to get carried away in the excitement of the moment.

It is simply a fact that the current analysis of a result might conceivably need to be recast at a later date after you have analyzed the results of future participants.

One of the trickier aspects of DNA testing as used by genealogists is that the number of variables you have to consider is very large indeed. What's more, many of them are not strictly quantifiable and many of those that will one day be quantified haven't yet been nailed down.

Most of us have an instinctive feel for what statisticians call 'significance' because without it we'd find making even simple everyday decisions very difficult indeed. But explaining DNA results to your participants requires you to describe significance in a particular way.

What are the factors?

There are no fewer than 11 factors that you'll end up considering when making a judgement about how closely two Y-chromosome haplotype results match each other. This goes some way towards explaining why study organizers end up using a qualitative rule-of-thumb approach rather than looking for a strictly quantitative formula and a black-and-white answer! The variables are:

1. Haplotype fit – the number of markers where a difference in match is found.
2. Number of steps' difference for each marker where a difference is found.
3. Mutation rates – for the individual markers where a difference is found.
4. Test resolution – the number of markers being compared.
5. Haplotype frequency – in the local population(s) of the participants.
6. Haplotype distribution – in continental and global populations.
7. Surname fit – degree to which the participants' surnames are identical or linked.
8. Surname frequency – in the general population.
9. Test participant selection method – were the participants selected randomly; and if not, how was this process weighted or influenced?
10. Surname categorization – the type of surname (as outlined in Chapter 3).
11. Family tree – is there a thesis being tested, or documentary evidence that links these two participants together?

To get to grip with the issues that this lengthy list raises let's consider some hypothetical examples.

Thinking about surnames

If two randomly selected Englishmen find that they have identical Y-chromosome haplotype signatures, this fact is not very significant in itself at all. The most common low-resolution haplotypes are shared by tens if not hundreds of thousands of men across the country. If these two men have the same very common surname such as Smith or Jones, it is possible that they may share a common male ancestor. But given that these surnames are so common in England, the fact that they share the same haplotype and the same common surname is only an indication that they might be related, it certainly is not proof of it.

However, if the surname shared by the two men is relatively unusual – and remembering from Chapter 3 that most surnames are rare – we'd be more likely to view their identical results as suggesting that they share a common male ancestor and that they belong in the same family tree.

Few people reading the last two paragraphs would have many problems with the broad conclusions drawn, yet its analysis used just 4 of the 11 factors I've previously listed, only 2 of which relate to the actual DNA result.

Thinking about test resolutions

This is the simplest variable to think about, but it is also one of the most contentious issues in genetic genealogy today. Using the same example as above where the two men have identical haplotypes, it's easy to comprehend that an exact match on 43 markers is much more significant than a match on 9 or 25 markers. The test companies that sell low-resolution tests with 9 to 12 markers are adamant that these tests are useful tools for genealogists. The problem is that many test organizers, including myself, who have retested a group of men with identical low-resolution results, have found that some of them are subsequently revealed to be completely unrelated to each other when the high-resolution test results are compared.

This debate is going to run for some time, but for now it's a good idea to direct new participants in your project at a minimum to take a medium-resolution test (24 to 25 markers), or to suggest that they take a low-resolution test as a way of joining the project (providing that the test company that you are using will allow you to upgrade to a higher resolution later).

Thinking about mutations

Now let us suppose that the haplotype results of the two men are not exactly identical. On one marker the values found are 12 and 13 while on a second marker they are 15 and 17. The differences are the result of accumulated mutations over long periods of time, so we recognize that a difference found on two markers is more significant than if just a single marker is different.

The working assumption is that a single-step mutation – e.g. the difference between 12 and 13 – is the normal mutational change that a marker undergoes and that a two-step difference – e.g. between 15 and 17 – can be the record of two single-step mutations or of a single double-step mutation. What this means is that you can generally see a two-step mutational difference as an indication of an earlier divergence from the common ancestor than you would when viewing a one-step difference.

Several other factors will influence the way you read the degree of difference between the two results. Firstly, a two-marker difference in haplotype results where 37 markers are tested is very different from a two-marker difference where just 12 markers have been tested. The resolution of the DNA test is always an important factor when comparing and interpreting results.

Secondly, since some markers mutate at a faster rate than others, if the two markers where the difference has been noted are thought to be ones where mutations occur more frequently – i.e. they are relatively fast-mutating markers – that would incline you to think that the two men are more likely to be related through a common ancestor than if the differences are found at two markers that are relatively slow to mutate.

Thirdly, some haplotypes (at whatever resolution) are more common than others in the populations of different continents. A review of the YHRD database of European haplotypes, for example, will indicate whether either of the two DNA signatures being compared has an unusual pattern of distribution in Europe or is simply rare across the continent. If both haplotypes are common and uniformly distributed then we would continue to question their degree of linkage. However, if both haplotypes are rare and are distributed in a similar and unusual pattern that would increase the odds that they are linked together despite their observed mutational differences.

Thinking about family trees
A final set of considerations would be to look at the two participants' surnames, their family history, and the circumstances in which they took part in the test.

As we saw in Chapter 3, many surnames originated in specific locations and still exhibit a strong regional distribution. If the two participants randomly decided to join the DNA study and live, for example, in Kalamazoo in America's Midwest and in the small Northumbrian village of Cramlington in north-east England, we would not assume that they need necessarily be related. However, if the surname they share is regionally strong in the north-east of England and both participants live within a few miles of each other in that heartland region, we might be more willing to consider that they could be directly related.

If their shared surname is one that might have had a single-ancestor origin, e.g. it is a locative type surname or one that is not common, then that might be seen as improving the odds that a suggested link does exist.

Finally, if their own researched family histories suggest that they could be related several hundred years' ago I would be much more inclined to accept the hypothesis that their DNA results demonstrate linkage than to reject it.

The interpretation process

Note that with the three last-mentioned variables our analysis is arguing back towards the DNA result rather than analyzing the result itself, and there's certainly a danger that one could end up in a circular argument and end up 'finding' a linkage between the two DNA results because you are expecting to find it. One way to counter this problem is to do all you analysis blind; another would be to set up all your hypotheses in advance. Family history, however, often advances itself by chance connections among different documentary files, so one has also to allow the DNA data to suggest new ideas. Generally, it is best to have a prior thesis that you are setting out to confirm or deny, but using circumstantial evidence is an important part of the interpretation process and there's no easy way to avoid this danger. You do need to review all the variables when framing your answer to the question posed by the two participants, but most of all you need to be aware that you're doing it. In the end you're looking for an analysis that is logically consistent, bearing in mind that this might not necessarily make it correct!

Reading over these hypothetical cases in the last few pages I'm sure that you're already putting them into an order of importance in your own mind. In general, it is important that your analysis and interpretation of the results should always start by looking at the actual result data and then applying a context to it. So my suggestion is that you frame your analysis by looking at the 11 variables I've identified in the following order:

1. The resolution of the test(s) being compared.
2. The different mutation issues.
3. The distribution of the haplotype(s).
4. The surnames of the participants.
5. Background information relating to the participants and their tests.

In the remainder of this chapter we'll look at ways in which you can quantify the degree of relatedness between two DNA test results and then I'll define a basic rule of thumb that you'll be able to explain to your test participants so that they can understand the likelihood of their being related. But before we do that, we need to look again at issues we approached in Chapters 3 and 4: surnames and mutation rates.

New developments

Looking at the list of 11 variables earlier in this chapter it will be clear that some of them do not relate to the actual individual test results themselves but to our collective knowledge about how markers and mutations behave. While the mutation rates of individual markers have not yet been quantified, this and other key DNA testing data will become much clearer as the number of men taking DNA tests increases.

As of mid-2004 some 25,000 people have taken at least one commercial genealogical Y-chromosome test. Once the online databases have grown by, say, a factor of 10 or 20 it will surely be clear not only which markers mutate rapidly but at what rate.

The current rule of thumb for the average rate of mutation for the average marker used by geneticists is that a single-step mutation will occur once in every 500 transmissions of the DNA. This implies that if you tested a group of 50 men who were all eleventh-generation descendants of a first-generation ancestor, on average you would expect to see one mutation per marker in the line of one of those 50 men. At this point it will be obvious why it is better to use high-resolution tests that compare 37 or 43 markers rather than low-resolution tests comparing 9 or 12 markers. Any differences between two compared haplotypes are much more likely to become visible from a high-resolution test and similar results are more firmly confirmed as such.

Several testing groups are now working on projects to quantify the mutation rates for particular markers, and these figures will have an important influence on the way genealogists use Y-chromosome testing once they are known. In the interim, Family Tree DNA has done a quick analysis of their clients' results and concluded that 15 of the 37 markers in their highest-resolution test are relatively fast mutating.

The study of surnames, as introduced in Chapter 3, is also undergoing change; in fact it is in the process of being revitalized due to the input from the first wave of Y-chromosome results. What we're going to see emerging is a new taxonomy of surnames, taking it beyond its etymological beginnings to include a deeper analysis of surname distributions over the centuries and the pattern of DNA results associated with each surname. This development is a key reason why I think genealogists should promote surname-based DNA testing, and I expect we'll all be benefiting from these developments quite soon.

I think too that there's a realistic chance that some genealogists working on low-frequency surnames may be able to produce complete or almost-complete genealogies for all the holders of these surnames. These comprehensive family trees would then become excellent tools both for checking mutation rates and to start classifying surnames by the pattern of their DNA results.

Five issues of interpretation

One issue that I've side-stepped until now is the simple question: how can one assume that the most common or modal Y-chromosome result is the original or oldest haplotype of a surname? The plain answer is that you can't, and as genealogists we shouldn't assume that it is so. Every family historian will recognize the plausibility of the following scenarios:

1. A family living in the area traditionally associated with their surname turns out to belong to a family tree that appears to

originate at the other end of the country.

2. Two families of similar size in, say, the eighteenth century suffer different fates whereby one dies out while the other expands with each generation.

3. An emigrant's family grows so fast that its descendants out number all the name-bearers from the original host country.

All of these are in fact examples taken from the Pomeroy scrapbook. In this surname's case the family with the oldest documented tree, the descendants of the original Norman family, are today fewer in number than the descendants of my own immediate family since the 1850s. In short, just as there's no direct correlation between the age of a family tree and the number of males who are alive today who can be tested, so one cannot assume that the modal haplotype is the core or founding haplotype for a surname.

To identify accurately the founding haplotype one would need a genealogy of the entire surname in much the same way as we saw in Chapter 3 that the study of surnames now looks to establish the genealogy of the founding line in order to define the surname's point of geographical origin and to corroborate its etymology.

A second issue is: at what point can one look at the distribution of haplotypes and decide that it indicates the surname has a single rather than multiple ancestors? Looking at the surname-based examples mentioned in Chapter 6, it's clear that a rare surname such as Mumma is more likely to be single-origin than not. Conversely the Pomeroy study has no fewer than seven low-resolution haplotypes that are common enough to suggest that these families pre-date surname formation, so this name clearly isn't of single origin. Deciding the answer for those surnames with results that fall between these two extremes is largely a matter of eye. Again, as more results are collated in the next few years a rule of thumb to define this will likely arise to guide our analysis.

A third point is that you can never assume a line of descent based upon a Y-chromosome result. As an example, let's return to the MacDonald study I mentioned in Chapter 8 and the DNA signature of King Somerled who died in 1164. Even if the chiefs of the MacDonalds can trace their ancestry back to King Somerled, the fact that someone else has the identical DNA signature does not mean that they could do so too. It only proves that they, and King Somerled, probably had a male ancestor in common. It is the chiefs' documented family tree that demonstrates the relationship with the ancient king, not the DNA result on its own.

Another way of thinking about this is to assume that King Somerled's DNA is not unique, that it was also shared by a number of his followers around his campfire. All their descendants could have the same DNA signature, but not all of them would be descended from a king. Distant cousins are not the same thing at all!

A fourth issue concerns sample sizes and sampling bias. In general one would want in a surname study to use a wide and random sample of male DNA. However, a study could conceivably be weighted to focus on the presumed original location of the surname or to match the present-day distribution of the general population (which is not the same thing as that surname's modern-day distribution).

Samples can be skewed for many reasons at any point in the selection process. Who is to say whether people with an oral tradition of their medieval origins are more or less likely to respond to your request to take a DNA test as a part of a surname study than someone who suspects that their ancestor was illegitimate or a naturalized immigrant?

A fifth issue concerns the use of statistical controls. These could – some would say should – be used to compare the results returned by individual DNA surname studies.

At a minimum as a study organizer I would like to compare the results for my British surname with a larger control group representing the population of these islands as a whole. Ideally one would want to have a control group representing the haplotypes found in particular geographical regions, in my case the English West Country, where my surname is believed to originate.

If you read more about the academic Y-chromosome studies of British populations you'll quickly come across the mention of the Atlantic Modal Haplotype, or AMH for short. This is simply the haplotype that combines the modal results found in a range of these British-population studies.

At the moment this is used as a shorthand reference point when comparing different results, but it marks the beginning of a process that will eventually be able to tell me which geographical region my haplotype is most firmly linked to and provide me with a list of haplotypes common in that region. The YHRD database is already moving in this direction.

Calculating the date of the common ancestor

At the beginning of this chapter I cited the question that study participants always ask: are we related? Whether the answer within a genealogical time-frame is 'yes' or 'no', it is possible to calculate roughly when
they might have shared their common ancestor in the male line.

This calculation, styled as the Most Recent Common Ancestor
(MRCA) formula, determines the number of generations that are likely
to have elapsed since the time when the two participants shared a common male ancestor. Unfortunately, this is one calculation where the
answer is expressed in terms of probability, and in practice this makes it
difficult for family historians to use.

For example, the formula might indicate that on a 37-marker test of
two men with a single-step difference on one marker that there is a 50%
likelihood that their common male ancestor was shared within the past 12
generations and a 90% likelihood that he lived within 27 generations. The
generation ranges widen further if a low-resolution 12-marker test is used.
Given that the date range of 12 to 27 generations could put the common
ancestor anywhere roughly in the period from 1325 to 1700, it's not hard
to see how relatively unimportant the formula is for genealogists.

The best use of this calculation is to use the range it sets to signal to
you whether there is a chance that you could one day document the link
suggested, i.e. whether the range includes a date close to or later than the
year 1500. The calculation also works well in cases where you're assess-
ing the potential links between ancient families, e.g. two clans who
believe they shared a common ancestor 1,500 years ago.

Note that the MRCA formula is built upon the average marker mutation rate mentioned earlier in the chapter. When this is refined and made
marker-specific then a revised formula may be devised that could be
much more useful. A free MRCA calculator can be found online from the
website accompanying this book with links to web pages that explain the
statistics behind the calculation.

Rule-of-thumb assessments

Genealogists are used to dealing with fuzzy situations and we generally
feel comfortable with imprecise answers to demanding questions. In

response to the question I outlined at the beginning of the chapter I'm quite happy to reply, 'Well, on balance it looks more likely than not that you two are related' or, 'Given this result, I doubt very much that you two are related.'

Common ancestry rule of thumb

If there's no exact measure of correspondence that we can use to define the degree of similarity between two Y-chromosome results and if the values created by the MRCA calculation are too broad, what rule of thumb can we use instead? Table 5 represents a set of useful guidelines rather than definitive statements: for what it is worth, this is the set that I'm using in summer 2004. Some of the assumptions underlying the table are different to those used by the testing companies who allow larger mutation differences. Notes on these issues are detailed on this book's website.

Some DNA study organizers might be more conservative in their reports to their participants whilst others might feel happier emphasizing the possibilities that remain open rather than those ruled out. That's a matter of personal preference. My own feeling is that I'd rather err on the side of caution and not inflate people's expectations. DNA testing is not a cure-all solution.

Table 5 Common ancestry rule of thumb

	Low resolution (9–10 markers)	High resolution (37–43 markers)
Identical Y-chromosome results	Potential link	Link highly likely
1-step difference at 1 marker	Potential link	Link quite likely
2-step difference at 1 marker	Link very unlikely	Potential link
1-step difference at 2 markers	Link very unlikely	Potential link
1-step difference at 3 markers	No Link	Link unlikely
3-step difference at 1 marker	No Link	Link very unlikely

It's clear that many people are misguided at present. I regularly get emails from men who have a different surname to mine asking me for details of my family tree so that we can see 'how we are related'. It is the duty of every surname study organizer to remind people that two people with unrelated surnames will virtually never find a documentable family tree connection even if their Y-chromosome haplotype is identical regardless of the number of markers that are tested.

Having looked at the ways in which you can analyze and interpret your study's results, in the next chapter I'll look at the ways in which you can present them to your participants.

More info at <www.DNAandFamilyHistory.com>

- Mutation rates per marker

- Online Most Recent Common Ancestor (MRCA) calculator

- Changes to the common ancestry rule of thumb

Chapter summary

- There are 11 variables you need to consider when analyzing Y-chromosome results

- The resolution of the tests being compared has always to be considered in tandem with the pattern of differences per marker

- As the mutation rates of individual markers become known it will become easier to make more precise estimates of how long ago two men shared a common ancestor

- Don't be afraid to be cautious when interpreting your results!

CHAPTER 13

How to Present and Publish
Your Study

Reaching your audiences

Writing up your DNA study's results, and keeping them updated as the project progresses, is a major task for a group test organizer. Not only must you assume that family members reading your report will have no background knowledge of genetic testing, you also have two audiences to address simultaneously: the internal audience of your own group of genealogy researchers and a global audience who are reading your study results online in order to compare them with their own.

If you look at the growing number of websites where group test organizers have presented their results and analysis you'll notice that it is difficult to convey this kind of complex data in an easy-to-understand format. Tables and diagrams can be worth a thousand words, but complex tables and diagrams are not always the answer in themselves. You also want people reading them to share the emotional excitement of discovery whilst providing well-documented research data to back up your findings.

What I recommend in this chapter is not the only solution to this problem, but I think it is the simplest one that the majority of people will find that they are able to adopt regardless of how much graphic design and website development experience you have. Even so, feel free to pick and mix ideas I mention here. Clearly, if you have access to high-powered graphics software you'll be able to go much further than I suggest here, and to inspire you I've mentioned on the website that accompanies this book those online presentations that I think are models of expositional clarity.

Many DNA studies use only very limited genealogical details supplied by each participant and make no reference to any documented family trees built up by researchers over the years that the study might be trying to link together. The method I outline here works for this basic kind of DNA study, but is equally applicable to those studies that have also built up well-documented family trees.

To make this chapter easier to understand I've included some examples from my own surname study. However, for reasons of space I've reduced the amount of data in the tables as I only want here to give an indication of the way in which the data should be organized, not to present my results. You can access the entire report on the Pomeroy study, with regularly updated tables and with explanatory notes, from the website accompanying this book.

Separating results from analysis

As you have twin roles as the guardian of your study's results and as the study organizer, it's a good idea to separate the two tasks that you face when presenting your study's results.

The first task is to present the results in a format that allows easy comparison with other DNA test programmes. The second and more demanding task is to interpret them and to explain their significance both to your participants and to a wider audience.

During the course of your study you've probably collected more data from the participants than you may decide in the end to publish, but make it a point early on to ensure that you have permission from the participants to publish the data that they've submitted to you (or data about them which you have researched yourself). It is best practice to make anonymous any data that you don't have permission to publish.

The two basic sets of data to be presented are:

1. The ancestry, lineage and background data of the DNA test participants.
2. The DNA test results.

While you can try to present these two sets of data together from the outset, in practice you'll find it difficult to make your analysis easily understood, or to update or check it, unless you separate them at the start. Some DNA project websites have an almost psychedelic intensity to them and quickly reveal what a tough challenge it is to try to cram multiple layers of data into a single two-dimensional table.

Background participant data

There is a set of ancestry data that it is useful to associate with each DNA test participant. These data collate:

1. Their full given names and surname.
2. The location where they live.
3. Their birthplace.
4. The birthplace and the birth year of their oldest known paternal-line ancestor.

Table 6 Sample table: participant data from the Pomeroy study

Participant ID	Fore-names	Surname	Residence	County / country	Birth-place	County / country	Ancestor's birthplace	County / country	Birth year	Family tree ID
1	•••	Pomroy	•••	•••	•••	Hampshire	Wilton	Wiltshire	c.1670	W1
2	•••	Pomeroy	•••	•••	•••	Hampshire	Launceston	Cornwall	c.1700	C1
3	•••	Pomeroy	•••	•••	•••	Durham	Liskeard	Cornwall	c.1630	C2
4	•••	Pomeroy	•••	•••	•••	Devon	unknown	Dorset	1680	D1
5	•••	Pomeroy	•••	•••	•••	Ireland	unknown	Ireland	c.1750	IRL1
6	•••	Pomeroy	•••	•••	•••	London	unknown	Canada	c.1920	CAN1

This sample table showing six participants is a subset of the full study. Note that the data have been anonymized. Ancestral data are taken from researchers' family trees. The two sets of ID numbers are set up by the DNA study organizer.

Of course, there is a wealth of additional data that you may choose to collect for each generation of your participants' paternal ancestry beyond the minimum given in Table 6, and you will certainly find it useful as your study develops.

Note that in the table the locations noted are broad rather than exact as this is the level at which comparisons can easily be made between participants. In the UK context this is the level of the county or region – e.g.

Cornwall, or the West Country – equivalent in North America to the state or province. I've anonymized the identity of the test participants for the purposes of publishing them in this book, but this can be a useful exercise in own right as it forces you to make the initial set of linkages based purely upon the DNA evidence in front of you.

If you can collect the details of each generation of the paternal ancestry from the study participant back to their oldest known paternal-line ancestor then so much the better. It is especially useful to collect data for each male in the family tree that includes the place and year of their birth and their mother's maiden name. When you are presenting the results of a large family tree where several members have taken part in your DNA study, the mother's maiden names will come in very useful. However, do not under any circumstances publish the mother's maiden name of either your DNA study participants or any ancestor of theirs who is still alive (foolishly, many banks still ask for the mother's maiden name as a security question when checking callers' identities).

Some participants may send their ancestry and lineage information to you in the standard electronic format known as a Gedcom file. This software allows family trees to be printed in reports that show the descendants or ancestors of particular tree members. These tree data will need to be stripped down by you to separate out the lineage data relevant for the DNA analysis, namely the male family members who either did pass, or could have passed, their genes down through the Y-chromosome.

Background family data

In addition to the data on each participant there is a range of data that you want to build up on each family tree. These include:

- The number of DNA test participants in the family tree.
- The total number of name-bearers in this family's family tree.
- The number of males in this family tree, subdivided into all males, living males and living adult males.

As your Y-chromosome study develops you will find that more than one man from the same documented family tree will take a DNA test, so gradually some of the data that you originally collected for participants will come to be associated with 'genetic families' and family trees rather than individuals. It's a good idea to decide on a simple descriptive label and code ID for each tree as these are much easier for people to relate to and recall.

It's worth recording at this point in your study your own thoughts about the likely origin of the surname(s) in your DNA study. Is it so unusual that everyone in Britain bearing them could trace back to a single ancestor within the last 500–700 years? Your expectation on this question is one hypothesis that your DNA study is setting out to prove or disprove.

Background surname data

Background data on your surname will benefit both your analysis and your report to participants. You should be able to find online reliable estimates of the number of people alive today in England and Wales and the USA bearing your surname. The UK figures are based upon the National Health Service (NHS) Register while the US figure is derived from the 1990 national census. The British figure can be checked at **<www.taliesin-arlein.net/names/search.php>** while the US figure can be very roughly calculated by multiplying the frequency for the surname found in the 1990 census at **<www.census.gov/genealogy/names>** by 245 million.

As an example, the total figure for the four surnames in the extended Pomeroy DNA study from the UK NHS statistics is 2,338. This fits well with the number predicted by research in our genealogy group. Based on our figures between 70% and 75% of this total are adults and around 48% are males, so as an approximate rule of thumb you can calculate that roughly 35% of an NHS surname total will equal the number of living adult male name-bearers in England and Wales.

Presenting DNA test result data

DNA results should be laid out in a table format with the markers set horizontally across the top of the page and test participants vertically down the page. Table 7 is a cut-down version of the kind of table that would be generated by a major study. Note that the commercial testing companies have collectively used 50 different markers to date so some fields will be blank in your table.

Table 7 Sample table: DNA results of Pomeroy study participants

Participant ID	dys 19	dys 388	dys 389i	dys 389ii	dys 390	dys 391	dys 392	dys 393	dys 425	dys 426	Genetic family ID
1	14	12	13	29	24	10	13	13	12	11	C
2	14	12	13	29	24	10	13	13	12	10	B
3	14	12	13	29	25	11	13	13	12	13	F
4	15	12	13	29	25	10	14	13	12	12	M
5	14	12	13	29	24	10	13	13	12	13	E
6	14	12	14	30	24	10	13	13	12	13	J

Raw result data, using just 10 markers, for the six study participants shown in the table of background data (Table 6). Once the haplotype frequency table (Table 8) has been prepared, the ID for each genetic family can be entered against each haplotype result in this table.

At every stage be very sure to check and re-check your data. Make sure that you have transcribed everything into your own files accurately so as to avoid the dreaded 'clerical mutations'!

Manipulating the DNA data

With the participant, family and DNA result data now in place, the analysis process of sorting the results into groups of identical or close-to-identical haplotypes can begin. If you are not comfortable with setting up a database or spreadsheet to manipulate this data you'll want to rely on the facilities to do so offered by the three leading testing companies. The best option is to put the data into a database format, as the process of comparing results when some participants have 43 markers and others have 25, 12 or 9 is complicated and you'll find that you will want compare the data in many different ways.

You can compare your results at any resolution of markers that you choose. Start with the 9-marker set that is common to all the leading testing companies on a low-resolution test and the YHRD database. These markers are DYS 19, 389i, 389ii, 390, 391, 392, 393 and the paired markers 385a and 385b.

Record the number of identical matches and the identities of the participants and families that match each other. Continue this process at higher marker resolutions, gradually breaking down the low-resolution DNA signatures into smaller subgroups. You'll see that, even with a few participants, your results will tend to cluster into groups. Use the rule of thumb for comparing two Y-chromosome haplotypes laid out in Table 5 on page 132 to help you label these clusters.

Table 8 Sample haplotype frequency table from the Pomeroy DNA study

Genetic family ID	dys 19	dys 388	dys 389i	dys 389ii	dys 390	dys 391	dys 392	dys 393	dys 425	dys 426	Frequency found
A	14	10	13	29	24	10	13	13	12	13	1
B	14	12	13	29	24	10	13	13	12	10	1
C	14	12	13	29	24	10	13	13	12	11	2
D	14	12	13	29	24	10	13	13	12	12	1
E	14	12	13	29	24	10	13	13	12	13	4
F	14	12	13	29	24	11	13	13	12	12	3
G	14	12	13	29	24	11	13	13	12	13	1
H	14	12	13	29	24	11	13	14	12	11	1
I	14	12	13	29	24	11	13	14	12	12	3
J	14	12	14	30	24	10	13	13	12	13	1
K	14	12	14	30	24	11	13	13	12	12	1
L	15	12	13	29	24	10	13	13	12	12	1
M	15	12	13	29	25	10	14	13	12	12	1

Results of 21 Pomeroy study participants aggregated into 'genetic families', each of which is then given its own ID. The aggregation process, here shown with 10 markers, can be done at any level of resolution.

Mapping the haplotypes

It is a good idea to check each DNA signature that you find in your results with the online YHRD database in order to find out how rare the haplotype is and where such results are most commonly found today. This you'll only be able to do at the low resolution of 9 markers. Note against each distinct haplotype some notes about where in Europe it is found and secondly its frequency. I've laid out detailed examples from the Pomeroy study on this book's website to show how this works in practice.

Table 9 Modified haplotype table showing YHRD distribution data

Genetic family IDs	dys 19	dys 389i	dys 389ii	dys 390	dys 391	dys 392	dys 393	Frequency found	YHRD frequency	Notes on European distribution
A-E	14	13	29	24	10	13	13	9	338	Spain (not Basque area), London, Netherlands.
F-G	14	13	29	24	11	13	13	4	771	Spain (Barcelona), London, Germany, Albania.
H-I	14	13	29	24	11	13	14	4	52	Spain (Basque area). Not in Scandinavia.
J	14	14	30	24	10	13	13	1	68	All Spain, London. Not in Scandinavia.
K	14	14	30	24	11	13	13	1	195	All Spain. Not in Scandinavia.
L	15	13	29	24	10	13	13	1	56	Western Iberia, Germany. Not in UK, Scandinavia.
M	15	13	29	25	10	14	13	1	1	Single record in Estonia.

Geographical distribution data taken from the Y-Chromosome Haplotype Reference Database of European samples at <www.yhrd.org> using seven of its nine markers. Even at this low resolution interesting information is revealed. Note that some of the results found in the Pomeroy study are rare in the database.

Mutation tree charts

A mutation tree chart aims to combine haplotype data within the framework of a family tree. It works by simplifying the traditional family tree so that it shows only a sub-set of males in the tree and generally outlines only the specific transmission events that lead to the present-day study participants. The haplotype results are shown by colour-coding the test participants on the tree. Using this method it is visually quickly obvious, when there are multiple participants tested in a single family tree, whether the tree has been correctly reconstructed by its researchers.

Figure 9 is a simplified black-and-white version of a mutation tree chart to show how the results can suggest new avenues of documentary research. You can see a range of colour charts, where the effect is much clearer, using data from the Pomeroy study on this book's website.

The standard diagram used to represent the differences between different DNA haplotypes is called a phylogenetic chart. These can be drafted at any resolution of markers.

Figure 9 Simplified mutation tree chart

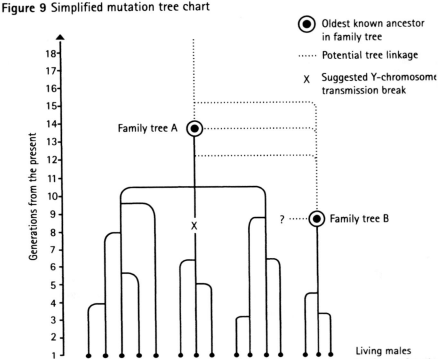

Members of Tree A exhibit two haplotypes with a strong modal result (W). The presence of haplotype Z in this well-documented tree suggests that there has been a break in the Y-chromosome transmission for some reason. The presence of haplotype W in Tree B suggests that a documented link will be found with Tree A, that Tree A is incorrect, or both.

You will need specialist software to produce phylogenetic charts accurately. There are several software packages that you can download to do this. These packages require you to install the software on your PC and to prepare and format your DNA test data to upload into it. These packages are not for the faint-hearted, so if you don't have a good relationship with your PC pause before attempting this exercise. I've laid out instructions on this book's website where you can find details of these packages.

You may decide once you've charted your phylogenetic haplotypes successfully that you want to associate other data with them. In my experience the easiest way to go about this is to recreate the chart in a graphics package rather than try to manipulate it using the original software.

In a phylogenetic chart each haplotype is represented by a circle, which is linked to the closest matching haplotype, based on the number of mutation steps between them, by a line. Each node in the diagram represents a single mutation step difference while the line between the nodes is labelled with the marker at which that mutational difference is recorded. The original software should emphasize those haplotypes where more than one result was recorded by increasing the diameter of the circle, but once you've recreated the diagram in a graphics package you have a choice of ways of doing this.

Figure 10 is a simplified black-and-white version of a phylogenetic chart using a subset of data from the Pomeroy study. You can see a range of colour charts, as adapted in a graphics program and where the effect is much clearer, on this book's website.

Figure 10 Phylogenetic chart showing Y-chromosome haplotypes and mutations

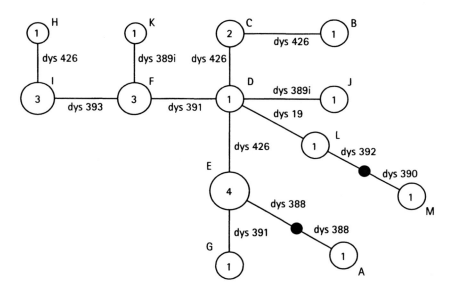

Note: Each node is a haplotype, each separation a single-step mutation. Haplotype labels correspond with the results in Table 8, with the number of participants in circles.

Among the data that you can associate with each haplotype in a phylogenetic chart are:

1. The number of family trees with this haplotype.
2. The total number of name-bearers, or number of males, or living adult males, in the family trees bearing this haplotype.
3. The geographical origins of those families.
4. The date of the oldest ancestor in each of those families.

You may get a clearer picture of your results by mapping just certain subsets of your data. For example, by charting only those families with their origin in a particular location you will end up with a regional haplotype distribution. You can see one example from the Pomeroy DNA study where I've isolated the Pomery results on the website that accompanies this book.

Geographical mapping

Geographical mapping of data can also emphasize aspects of the results with a great deal of clarity. You can map:

1. The present-day locations of the men taking the DNA test.
2. The locations where they were born.
3. The origin of their oldest-known paternal ancestor.

You can import maps into graphics software programs and then annotate and colour-code them to highlight the data you want to emphasize. You can see a set of maps from the Pomeroy study on this book's website.

Timeline charts

A timeline chart can be used for both single ancestor and whole surname studies. It plots the origins of known family trees, grouped according to haplotype, against a timeline running from medieval times – say, the year 1400 – to the present day. This allows you to compare the age of the different family trees within each haplotype and, by colour-coding the locations of origin for each family, to see how families of different origins are spread among the different haplotype clusters.

There is no software to generate these. The example on this book's website was created in a graphics program.

Reconstructing family trees

One of the key outcomes of a Y-chromosome study is that you should be able to define some new working hypotheses to assist people researching and reconstructing their family trees. To do this well you really need to have kept participants' lineage and DNA data separate, otherwise the ideas you will generate could feel to be something of a circular argument.

As your project progresses you'll want to show how the use of the DNA tests, which has stimulated new documentary research, has led to improvements in the integrity of the groups' documented family trees. If you record a snapshot of the state of your knowledge at the start of the project you will be able to show later how far you've improved your research.

The degree of improvement may well surprise you, particularly as you find that you are able to progressively move back the earliest date of the documented trees and to pinpoint their geographical origin. This type of information is also very useful for the regular reports you send to the DNA study participants and when you are trying to persuade other men to take part in your project.

It's important to stress that the results you gain from the DNA testing of a surname group will start an iterative process that involves checking the documented family trees, correcting mistakes, reviewing the results of further DNA tests, checking the trees, and so forth. Some of the errors that you are likely to uncover in the documented trees will have been embedded in them for years and they have probably long been accepted by all the concerned researchers. Challenging such long-held assumptions may raise eyebrows within the wider genealogy group, but the new hypotheses posed will over time help create a fully documented picture of the family trees within your surname group that is consistent with the body of DNA results you've collected. However fascinating a DNA study becomes, it is a good idea not to take your eyes off the ultimate goal of the genealogical project: to reconstruct a complete set of accurate, documented family trees.

Online documentation and study report

While it's hard to imagine a DNA test organizer without access to the web, perhaps only one half of us have created a website to promote our DNA study or to present our results. As I outlined in Chapter 11 when discussing ways of promoting a DNA project, even a simple website is a tremendous tool to market the study and explain its results and impact. If you don't want to build one yourself, some of the testing companies will provide you with web space to do just that.

The best approach to online publishing nowadays is not to set up a complicated, multi-level website. It's far better just to have a few web pages to explain the project and then to deliver the much longer and detailed information containing the results and your analysis in the format of a PDF document.

PDF software is very well established nowadays. Downloading the reader software is free from makers Adobe, and it is even possible for you to create your own PDF document for free by uploading a traditional document such as Microsoft Word on to the web for conversion into PDF format. You can easily email a digest of your results in PDF format to potential test group members to solicit them to join the programme as well as to current members of the study group to update them on progress. As your project develops, it is easier to change the underlying document in Microsoft Word than to update a bespoke web page. It also offers you more organizational flexibility. If you want at some stage to pass on the organization of your surname project to someone else, then someone without web design skills can take it over. What's more, fewer people have security fears when downloading a PDF file compared to, say, a Word document.

The final great advantage of a PDF document is that you can print it out on regular paper from any colour printer without further hassle. That means you can print it out yourself and send a copy by regular postal mail to people who do not have email without any further work for yourself. This is one of the great advantages of the digital age, a one-size document that fits all of your different purposes.

The best example on the web of a PDF summary report of a Y-chromosome study has been put online by Doug Mumma, one of the early adopters of DNA testing, detailing the results of his Mumma DNA study.

If you decide to go the website route the options to present, organize and navigate your data are far greater, but unless you make a living from web design you'll find that learning how to present complex results in even a simple tabular format is a far more difficult challenge than learning how to do it in MS Excel or MS Word and then converting into PDF format. That said, I've listed on this book's website those websites that I think do the best job in presenting their DNA project material and which I think will serve as an inspiration to you.

More info at <www.DNAandFamilyHistory.com>

• The best surname DNA study websites

• Online resources including surname frequencies, and charting and PDF software

• Pomeroy DNA study results

Chapter summary

• Separate the presentation of the DNA test results from your analysis of them

• Collect additional data about the participants' family trees in order to add an extra level of detail to your study

• Save your DNA study report as a PDF file to facilitate easy printing, emailing and downloading from your website

• Update your results periodically and use this updating process to promote your programme to potential participants

CHAPTER 14

The Future of Genetic Genealogy

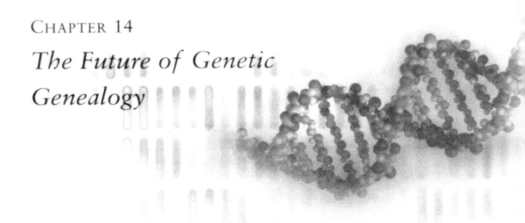

What will come next?

The short-term future of genetic genealogy looks very interesting indeed. The market is growing fast, in terms of both the number of companies offering DNA tests and number of people taking them. Online resources such as databases of results are growing in number, size and, more importantly, in terms of their complexity and usability. The first generation of data from the collective total of genealogical Y-chromosome test results is just beginning to be released in proper academic form. First up will be empirical data on the mutation rates of individual markers.

It's been a rapid ride since genealogical DNA testing began in the year 2000. Here are 10 trends that I expect to see become more prominent before the second edition of this book is published.

1. Growth in the number of DNA surname projects

Around 1,500 surname projects are currently under way worldwide, fewer than half of which were started before 2003.

Even so it's still early days. No more than 50 of these surname studies have more than 50 participants' results reported, which is probably the minimum number that a study of a relatively common surname needs in order to create a reliable DNA signature matrix for the surname as a whole. Many more studies will reach this level over the next year or two.

2. More 'deep ancestry' tests

In the near future SNP tests will be offered more widely as researchers demand to identify their deep ancestral haplogroup – as defined by the Y-Chromosome Consortium – as well as to investigate their genealogical history using DNA testing. Expect to see a new wave of DNA test advertising using the word 'roots'.

3. Low-resolution tests will be used less

Low-resolution Y-chromosome STR tests using 9–12 markers will quickly become less popular among surname groups as their limited value becomes increasingly apparent. What's more likely is that companies will continue to sell low-resolution tests to clients who are unaware of their limited use. Low-resolution tests currently on offer that attempt to label ethnic or ancestral origins – such as descent from Genghis Khan – may also become to be seen as requiring a similar improvement in terms of resolution.

The number of markers in the high-resolution tests is unlikely to increase much beyond 43 even though more than 200 potential STR markers have been identified on the Y-chromosome. Expect to see different configurations of markers offered in high-resolution tests, particularly when useful data are released about their individual mutation rates. There's a limit to how far the configuration of markers can be changed, however, as the 'installed base' of genealogy clients with DNA test results that they want to compare is growing all the time.

Test companies are likely to offer a wider range of retesting options that allow existing customers to add new markers to their overall DNA signature. People will start to get more used to the idea of archiving their DNA test samples rather than destroying them in order to access cheaper retesting rates.

4. New types of test on offer

It is possible that the extraction of DNA from household materials belonging from previous generations may become feasible in a test

marketed at genealogists. This could lead to the marketing of DNA tests
that extract DNA from old artefacts including letters and clothing.

5. Quantified marker mutation rates published

Calculations used in genetic genealogy are built at present around a number of numerical assumptions, the most important of which is that mutations occur in the average STR-type marker roughly 1 in every 330–500
generational transmissions. It's known that each marker mutates at a
slightly different rate, but not how varied those rates are or the specific
rates for each marker.

At least one study is under way, sponsored by Family Tree DNA and
using inputs from some of their surname DNA projects, to quantify the
mutation rate values for individual markers. As this type of data becomes
available, expect the testing companies to change the panel of markers
they offer or to move to an à la carte system of marker selection.

6. Improved online results' databases

The smarter development of more useful online databases is an exciting
prospect. At present the testing firms' databases are light on the kind of
background information that genealogists and surname analysts would
find useful to enable the subdivision of the results in the database into
more tightly focused control groups – for example, built around regional
or clan identities. The next generation of reference database will probably
develop in this direction as the quality of the results database becomes
one of the key ways in which the testing companies can differentiate
themselves in an increasingly active market.

Improvements in nomenclature and cross-referencing results' data
from different companies should make databases easier to use.

7. Regional haplotype data published

One offshoot from the improvement in the results' databases will be
the publication of data relating to regionally defined DNA signature

haplotypes, as well as data about the regional distribution of specific haplotypes and individual marker values.

This could lead to the identification of more secure associations on the one hand between specific haplotypes and ancestral groups such as clans and on the other hand associating surnames with distinct geographical areas. Expect to see more clan-type studies getting under way as regionalized DNA result data becomes available as well as the increased use of comparative DNA results within every type of DNA project.

8. Technical improvements in the 'wet science'

There are many technical issues in the lab relating to the complex, multi-stage testing process – colloquially known as the 'wet science' – that are in the process of being improved.

The push to improve the refinement of marker sets will continue. New STR markers and the primer kits to analyze them are being classified every year, so this could drive changes in the selection of markers for the higher-resolution tests.

Inside the lab itself the physical testing machines are getting bigger and faster, and it is gradually becoming easier to perform quicker tests by improving the testing process.

Calibration issues have been a major issue for testing companies in the past. There are international standards set to ensure that DNA results can be compared from lab to lab (one reason why the YHRD database, which contains results derived from calibrated and quality controlled labs from all over the world, is such a useful online tool). Expect to see these quality issues pushed by the testing companies more and more to promote themselves.

9. New software

The growth of genetic genealogy and its movement into the mainstream could stimulate the development of a range of new consumer-friendly charting software. This is needed to bridge the gap between traditional Gedcom-friendly family tree software and the specialized presentation of Y-chromosome results data that DNA study organizers want.

10. Widening impact on history studies

On the statistics side, as more data is collected from genetic genealogy projects, the way is open to revisit one of the most debated statistics in population history: what is the true rate of illegitimacy, and does it vary by region, historical period, or any other variable? It's possible that the historical illegitimacy rate in western Europe will turn out to have been higher than is anticipated, a feature that could dramatically reduce the usefulness of DNA testing for genealogists.

More widely, the acceptance of DNA testing by the genealogical community could be the start of the development of a new area of historical research, one which uses the increasingly widespread availability of genetically confirmed genealogies to review, and potentially rewrite aspects of social history and surname studies. In the next few years we should be able to create the first classification of surnames that unites etymological, genealogical and genetic research to define which types of surnames appear more likely to have a single-ancestor rather than a multiple-ancestor origin. We'll certainly be looking closely to see, for example, whether the DNA results for topographical-type surnames appear to have a single common ancestor. The growing numbers of surnames for which we have a genetic profile could turn our thoughts on surname origin and transmission upside down.

The last word

There's never been a better time to be researching one's family history. Mass transcription projects are placing huge amounts of data online every month on a free-to-view, pay-per-view or subscription basis. In Britain, constructing one's family tree back two centuries is now relatively easy, and I expect in time it will be possible to view on the web the detailed reports of researchers who have been able to reconstruct entire surname trees to a great degree of accuracy and completion. DNA testing will come to be seen by researchers as an increasingly important tool when linking family trees prior to the period of modern documentary records.

And who knows, maybe it will come just in time. In Britain we are rapidly witnessing the breakdown of the centuries-old system whereby children automatically take on their father's surname. The breakdown of the surname transmission process is growing with each decade. The process is a gradual one, but genealogists in 100 years' time will surely be warning 'newbie' researchers that surnames had stopped being a reliable way of linking families together before the mid twenty-first century. At that point, DNA typing will truly have come into its own.

One thing's for sure: the DNA story is going to run and run. Even as I edit this final chapter in July 2004 I've recently become aware of a new academic paper published in the *American Journal of Human Genetics* by a team under Manfred Kayser at the Max Planck Institute of Evolutionary Genetics in Leipzig, Germany. The paper describes how they reviewed the entire Y-chromosome looking for new SNP markers: they found 166. Just a few days earlier an even more astonishing paper hit the mainstream press when a team from University of California at Santa Cruz announced that the so-called 'junk DNA' might have an a crucial evolutionary role after all. Amazingly many stretches of junk code very similar to that in humans are found in other mammals such as mice and rats. The fact that sections of this code has remained unchanged in all three species since our last common ancestor roughly 400 million years ago indicates that it may do an important job. As yet its function is still not clear.

As I send this book off to the publisher the world's largest surname test project – Orin Wells' global project for the Wells surname – stands at 303 participants. Impressive and inspirational as this achievement surely is, I expect that by the end of the decade a surname study of these dimensions will be commonplace. The largest projects then could number 1,000 participants or more.

I must confess that I'm greatly looking forward to reading the accounts of the next generation of DNA study pathfinders and reviewing the new and challenging questions they will inevitably throw up in their wake.

More info at <www.DNAandFamilyHistory.com>

- Genetic genealogy trend update

- Latest news for genealogists on the impact of DNA testing

- Sign up for my email DNA newsletter!

Chapter summary

• High-resolution Y-chromosome tests will oust low-resolution tests

• Online DNA result databases will dramatically improve in quality, scope and usefulness

• More DNA results will be aggregated, stimulating new quantified data on regional haplotypes, marker mutation rates and other variables

• Related study areas such as surname analysis and social history will start to benefit from inputs from genealogists who have been able to build accurate DNA-verified family trees

ACCOMPANYING WEBSITE

\<www.DNAandFamilyHistory.com\>
This book is just a beginning. On the website you can view a wide range of updated materials that expand upon the topics in this book including:

Organizing a DNA study

Latest DNA test prices

Comparison of the testing companies

Chris Pomery's email newsletter: trends and news for genetic genealogists

Online database of Y-chromosome surname studies

The DNA study organizer's checklist

DNA study etiquette: privacy, permissions and data protection for British genealogists

Sample documents: consent form, research form, emails to potential participants

Online resources for study organizers: forums, web rings, databases

Online marketing resources: email, charting and PDF creation software

Books and online reports that can help you prepare for your study

Results' interpretation

The original Sykes DNA study

A worked example: the Pomeroy DNA study

Analysis of the leading DNA surname studies worldwide

Analysis of the leading single-ancestor case studies

Analysis of the leading clan, caste and locality-based studies

Special markers

Mutation rates per marker

The common ancestry rule-of-thumb table

The online Most Recent Common Ancestor (MRCA) calculator

Contentious issues under debate, variables and other issues of interpretation

Haplotype resolutions

The 'Modal versus Multiple Origin' debate

Online results' databases and issues when inputting your own result data

Dislocations between 'genetic trees' and 'family trees'

Data on non-paternity rates

Results' presentation

The best surname DNA study websites and reports

Maps: haplotype distribution and other maps

Charts: mutation tree, phylogenetic, timeline and other charts

Genetics

Genetics for beginners

MtDNA and autosomal DNA tests, 'deep ancestry' tests

Notes on ethnic and location-oriented commercial tests

Forensic DNA tests and paternity tests

Glossary of terms used in genetics

Human migrations

Out of Africa: migration routes to Europe, Asia and the Americas

Explanation of the Y-Chromosome Consortium's chart of clades

Descriptions of the main Y-chromosome and mtDNA clades

Y-chromosome studies of British populations and the Atlantic
  Modal Haplotype

Development of statistics of British regional control populations

Interesting academic papers on migration studies

Surnames

Developments in the study of surnames

Online data including surname frequencies

Visit the website to:

Access DNA test prices and subscribe to an email newsletter about
  DNA testing.

Email feedback to me at **feedback@DNAandFamilyHistory.com.**

GLOSSARY

allele Within the context of Y-chromosome analysis, an allele is an item of genetic information within a gene. It is expressed numerically as an allele value. An allele changes through mutation, but only rarely.

allele value The count of the number of tandem repeats found at a particular STR marker (and sometimes adjusted according to agreed norms). In a genealogical DNA test it is almost invariably expressed as a whole number. Each value will fall within a tight range of values that are associated with each marker. Most DNA test results produce values that are highly concentrated towards the most common or the two most common values (i.e. close to, or like, a normal distribution).

Atlantic Modal Haplotype Name given to the most common haplotype that has been identified in large-scale British and European Y-chromosome testing projects.

autosomal DNA The DNA found in the 22 pairs of chromosomes (known as autosomes) other than the sex-determining 23rd pair (the sex chromosomes). Autosomal DNA is subject to recombination, or genetic shuffling, during the process when offspring are created. It is the focus of forensic genetic tests that seek to determine unique identities.

base pair DNA is composed of two strands that are made up of chains of four individual bases (roughly equivalent to nucleotides): adenine (A), thymine (T), cytosine (C) and guanine (G) each of which pairs up with its complementary base on the opposite strand (according to the rules A–T and C–G) to form a base pair. The linked strands create the double helix pattern now universally symbolically associated with DNA.

Beringia The now-submerged continent between Alaska and Siberia. This land bridge has been dry land during most of the past 2 million years though during the current interglacial period its centre lies drowned beneath the seas.

byname Prior to the adoption of hereditary surnames, people were distinguished by a second by-name as well as their forename. These bynames could be a nickname, a reference to parents, an occupation or a place name. Some over time became surnames.

Cambridge Reference Sequence Reference definition of the 16,568 base pairs of human mitochondrial DNA, first published in 1999 and revised last in 2001. See <www.mitomap.org>.

caste A social class defined by hereditary rank, profession or wealth.

cell The basic unit of life, 'a miniature factory producing the raw materials, energy and waste removal capabilities necessary to sustain life' (Butler, 2001). A cell consists of the cytoplasm, the body of the cell, and the nucleus which contains the chromosomes. The average human has around 100 trillion cells.

chromosome Made of DNA and special proteins, chromosomes are the receptacle for genes which are arranged linearly along them in sections known as coding regions. Humans have 23 pairs of chromosomes, one each from the mother and from the father. (This is the number in humans; other species have more or fewer pairs of chromosomes.) *See* autosomal DNA, sex chromosomes.

clade In the context of migration studies, a group of modern humans whose members share an evolutionary origin derived from a common ancestor. Synonym for haplogroup.

clan A grouping, based upon allegiance to a laird or chief, built up through geographical and kinship links. Some clan lineages are well defined by oral tradition within certain clans. The term is used to describe kinship groups in both Ireland and Scotland.

clerical mutation First defined by Doug Mumma to describe the type of false mutation caused by a project administrator or by scribal error. Prone to bedevil even the most carefully organized projects.

coding regions Sections of a chromosome containing genes. This is in contrast to the non-coding regions, or junk DNA.

Data Protection Act Data protection is governed by national law in European countries. In the UK this is covered by the 1998 Data Protection Act which outlines the principles, rules and procedures under which personal identifiable data may be held in the UK. British-based genealogists are exempted from the requirement to register under the act. *See* notes on <www.DNAandFamilyHistory.com>.

deep ancestry Ancestry prior to the standard genealogical time-frame, the period when hereditary surnames came into effect, and stretching back to mankind's origins in Africa.

DNA Deoxyribonucleic acid, the 'read-only memory of the genetic information system' (Passarge, 2001). DNA is the main component of the chromosomes that reside within each cell, and it contains the coded instructions required to replicate the cell and associated enzymes. Its structure is that of a double helix. Its ability to replicate itself accurately results from the system of the base pairs of which it is composed.

DNA signature This author's shorthand phrase to describe the haplotype that belongs to an individual DNA study participant.

DNA test Within a genealogical context, a multi-marker test on the Y-chromosome. Can also refer to mitochondrial DNA tests and autosomal DNA tests.

double helix Description of the structure of DNA. It resembles a twisted ladder, the rungs being the joined base pairs and the sides their associated sugar molecules.

double-step mutation A mutation in which two base pairs at one time are changed within an allele, resulting in a difference of two integers in the allele value when comparing two DNA test results or DNA signatures. *See* single-step mutation.

DYS Acronym for DNA/Y-chromosome/Single-copy sequence. Each DYS marker includes a number which indicates the locus number in sequence of discovery. Other locus descriptions also exist – e.g. GGAAT1B07 – which describe the actual location of the locus on the chromosome. These may, in time, be given an official DYS sequence number.

genes The discrete messages contained in the chromosomes that produce the proteins that have a vital or useful function in a human being. Humans have roughly 60,000–80,000 genes, only five to six times more than the humble fruit fly.

genetic bottleneck When a population (of humans or any other species) is subjected to sudden trauma or initiates a sudden change such as a migration, only a few members of the original population survive. As this surviving group breed and expand the distribution of haplotypes in the ensuing population will appear lop-sided and contain relatively less variation than a comparable unaffected population. *See* genetic drift.

genetic clan A shorthand phrase to describe members of a clan who share a common genetic heritage through identical, or near identical, haplotypes. Popular science writers also use the phrase to describe haplogoups, clades and genetic lineages, often giving them fictional identities.

genetic diversity The principle of migration studies that states that the older a population is the greater number of different haplotypes it will contain, i.e. the more genetic diversity it will have.

genetic drift The process in which the genetic diversity within a small population is reduced over many generations. This happens because the small size of the breeding population means that at each generation only a sample of the available different genotypes is passed on and so by chance some genotypes become greatly reduced in frequency or lost for ever. Genetic drift is accelerated when there is a genetic bottleneck. It is a random process, and should not be confused with genetic change due to natural selection, which operates all the time.

genetic family A loose shorthand phrase to define men that have an identical, or near identical, haplotype and who, in the context of this book, usually share a common cultural heritage or surname.

genetic refuge A location where humans, caught up in an external genetic bottleneck event such as profound climate change, survive until a subsequent change often allows them to re-expand back into these locations to reclaim them. The best example is Beringia, the now-submerged continent between Alaska and Siberia.

genetics The branch of biology that investigates how genes control processes including the mechanisms of hereditary transmission and the variation of inherited characteristics among related organisms.

genome The total amount of genetic information that an organism possesses, i.e. the sum of its genes and its non-coding DNA plus an account of its genetic architecture, e.g. how many chromosomes it has.

haplogroup Synonym for clade and defined by the differences seen in a specific set of SNP markers. Used to describe different groups of humans as defined by long-term migration histories.

haplotype A string of allele values for different markers, primarily STR markers, that collectively make up the result of a Y-chromosome test delivered by a testing lab. Haplotypes can consist of values of any number of markers. See resolution.

HUGO The Human Gene Nomenclature Committee, based at University College London, which defines the numbering of STR markers. *See also* NIST.

interglacial optima Regular brief warm and relatively wet climate periods that punctuate the colder drier periods known as Ice Ages.

interstadial Similar to an interglacial optimum but briefer.

junk DNA The non-coding sections of the chromosomes which have no known function. However, as they undergo mutation in the same manner as genes in the coding regions geneticists can track and classify them. Genealogical DNA tests use markers contained within the junk DNA of the Y-chromosome. Some 90% of the DNA in human chromosomes are defined as non-coding regions. However, the fact the much of this 'junk' DNA is essentially the same in humans as in our fairly remote relatives, the rodents, suggests that it may yet turn out to have a useful evolutionary role.

locus The position of a gene, or of a DNA marker in a non-coding region, is called a locus (pl. loci).

Last Glacial Maximum The greatest extent of the ice shield that spread across the northern hemisphere during the last Ice Age some 18,000 years ago. In Europe the ice reached as far as southern Britain and turned most lands north of the Alps into steppe-like tundra or polar desert.

marker A locus within a non-coding region where genetic variation has been identified. *See* SNP, STR .

mitochondrial DNA The mitochondrion is a small structure in the cell's cytoplasm that is often described as the cell's battery as it performs the essential function of creating, storing and transferring energy within the cell. It has its own DNA (mitochondrial DNA or mtDNA), independent of the DNA in the chromosomes in the cell's nucleus. The genetic code in mtDNA is passed from a mother to her children, both male and female (fathers cannot transmit mtDNA as it is contained in the egg, not in the sperm). Thus mtDNA travels down the generations only through mother-to-daughter lines.

modal A statistical term that describes the most commonly found value of a range of values.

Most Recent Common Ancestor The Most Recent Common Ancestor (MRCA) calculation determines the number of generations since two DNA haplotypes diverged from a common ancestor. Potentially this point could lie at any time since the first humans migrated from Africa. The calculation is highly inexact from the genealogist's point of view. The results of an MRCA calculation are expressed in terms of a percentage probability (usually 50% or 90%) plus the range of generations in which the result is likely to lie. The calculation is most useful in clan studies where their two-millennia time-frame is long enough to make the MRCA calculation useful. It is less useful for surname-oriented genealogists who are concerned with a much shorter time-frame.

mtDNA Abbreviation of mitochondrial DNA.

multi-copy markers A special marker where two or more copies of the same marker exist at different loci on the Y-chromosome. These markers are designed by a small letter suffix, e.g. DYS 385a and DYS 385b.

Multiregional Thesis Theory that is opposed to the 'Out of Africa' thesis and which holds that the species Homo sapiens arose in several places on the planet more or less contemporaneously and developed separately for most of mankind's history.

mutation DNA alters through mutation. This can happen in many ways, primarily though the substitution of one base pair by another or the deletion or insertion of one or more base pairs. Genealogical DNA tests examine the non-coding regions of the Y-chromosome and count the number of base pairs at specific loci to reveal mutations that are either deletions or insertions.

NIST The US-based National Institute of Standards and Technology, which publishes a 'federal DNA quality assurance standard that helps forensic and medical laboratories ensure that DNA profiles made by the fastest and most popular profiling method are accurate.'

non-coding region *See* junk DNA.

nuclear DNA The cell's nucleus is the core that controls the functions of the cell. The DNA contained within the nucleus is called nuclear DNA to distinguish it from mitochondrial DNA.

nucleotide A subunit of DNA; for practical purposes, equivalent to one base in a base pair. *See* SNP.

'Out of Africa' thesis Theory, now almost universally accepted, that mankind's origins lie in Africa. Debate still rages, however, whether the rest of the world was populated by a single migration from Africa or multiple migrations over an extended period of millennia. Built originally around archaeological and palaeoanthropological findings about extinct human species, the theory was given a huge boost by a ground-breaking paper published in *Nature* by Rebecca Cann in 1987 and subsequently by hundreds of other papers reporting mitochondrial DNA and Y-chromosome results. The opposing theory is the Multiregional thesis.

paired markers A special marker that produces a result with two allele values instead of one. Usually recorded divided by a comma or hyphen, e.g. marker YCAII 11–19.

patronymic A system of naming where the son takes his father's forename as a surname.

phylogenetic tree Tree chart showing the ancestor and descendant clade relationships, drawn using data from studies in phylogeography.

phylogeography The study of the history of the geographical dispersal of DNA molecules across the planet. This is different from classical population genetics which is the mathematical study of the history of whole human populations (see Oppenheimer, 2003). The phylogeographer's most appealing output is the geographical mapping of the human clade chart.

polymorphism The situation where there exists in the population more than one DNA allele at a given locus. For the purposes of Y-chromosome analysis, there are two types of polymorphism: sequence polymorphisms and length polymorphisms. SNPs are sequence polymorphisms while the STR mutations tracked in genealogical Y-chromosome tests are length polymorphisms.

protein Proteins, each of which is made up of a specific sequence of amino acids, perform the biochemical reactions required to maintain the life of the cell. The genetic messages inside our DNA are actually the code required to generate all the different proteins needed for that specific cell.

recombination Recombination is an elegant and efficient method for the genome to experiment with gene adaptation without leaving this solely up to chance mutation. Each person's set of chromosomes comprises in every pair one inherited from their mother and one from their father. During the process in which eggs (in a woman) or sperm (in a man) are produced, the chromosome pairs exchange stretches of DNA with one another, which are then, in the new 'shuffled' form, passed down to the new offspring. From the genealogist's point of view recombination is a problem as it effectively jumbles up the offspring's DNA message. The DNA tests used by genealogists target the Y-chromosome which does not undergo recombination during the production of sperm.

resolution Every haplotype is defined by its degree of resolution, i.e. the number of markers that were measured. A 4-marker haplotype is clearly not as well defined as a 43-marker haplotype. Genealogists should avoid low-resolution Y-chromosome tests whenever possible and at a minimum choose a medium-resolution test of 24 or more markers.

sex chromosomes The sex of an individual is determined by the 23rd pair of chromosomes. Females have two X-chromosomes, while males have an X-chromosome and a Y-chromosome. A man inherits one X-chromosome from his mother (which is at random either the one that came from her mother or from her father), and inherits the Y-chromosome from his father.

significance A statistical term and as such to be used by non-statisticians only with care. Broadly, it gives an estimate of whether a statistical result has a genuine underlying cause or whether it has arisen by chance. Genetic genealogists will come across it primarily when reviewing MRCA calculation results which are expressed in terms of different confidence levels. However, it also encompasses the entire art of describing to two project participants whether their results match or not.

single nucleotide polymorphism *See* SNP.

single-step mutation A difference of a single integer in the allele value at a single locus when comparing two DNA test results or DNA signatures. It is by far the most common type of mutation that a genealogist will encounter: 'The majority of STR mutations involve the gain or loss of a single repeat unit' (Butler, 2001).

SNP Abbreviation of 'single nucleotide polymorphism'. This type of polymorphism takes place at only a single nucleotide at a specific position in a particular chromosome. Most importantly, this change occurs only once in a single individual. SNPs are like an on/off switch that was thrown at a particular moment in time and that has not changed since. When you are DNA tested on an SNP marker you will have either a result A or result B. The group of descendants that each SNP defines is known as a clade or haplogroup. SNPs are common and can be population-specific, i.e. the identity of the geographical (and possible ethnic) origin of the participant can often be revealed.

STR Abbreviation of 'short tandem repeat'. In this type of polymorphism a very short sequence of DNA in a non-coding region is repeated a specific number of times (the allele value). Test companies like them because they can measure several STRs at the same time in their lab processes. Geneticists like them because there are many of them and some are fast-mutating while others are slow to mutate. The standard Y-chromosome tests sold to genealogists measure STR markers.

tandem repeat *See* STR, VNTR.

tribe A number of families, clans, or other groups who share a common ancestry and culture, in most cases associated with a particular geographical region. DNA tests generally set out to link people in two ways; firstly, by identifying the geographical areas where similar results have already been found; and secondly, by attempting to pinpoint a clan affiliation for similar or identical samples. While the first process is well established the second is still in its infancy.

VNTR Abbreviation of 'variable number of tandem repeats', a type of polymorphism in which stretches of non-coding DNA carry DNA sequences that are repeated a varying number of times. In this book I've instead used the phrase STR markers, which technically are microsatellite VNTRs, as opposed to longer tandem repeat markers which are always described as minisatellite VNTRs.

Y-chromosome The sex chromosome that is present only in males. Mimics the father-to-son transmission of surnames as its DNA is handed down without the shuffling of recombination. Surname-based DNA studies should used only Y-chromosome DNA tests and male participants.

Y-Chromosome Consortium Collaborative effort by geneticists to create a flexible descriptive framework for Y-chromosome clades. Its results were defined in a paper published in *Genome Research* in 2002 (see Further Reading). It periodically upgrades its work at <http://ycc.biosci.arizona.edu>.

FURTHER READING

J. Butler, *Forensic DNA Typing* (Academic Press, 2001)

K. Cameron (ed.), *English Place Names* (Batsford, 1996)

L. L. Cavalli-Sforza *et al., The History and Geography of Human Genes* (Princeton University Press, 1995)

L. L. Cavalli-Sforza, *The Great Human Diasporas* (Perseus Books, 1995)

L. L. Cavalli-Sforza, *Genes, People and Languages* (Penguin, 2001)

A. Chamberlain and M. Pearson, *Earthly Remains: The History and Science of Preserved Human Bodies* (British Museum Press, 2001)

B. Cunliffe, *The Ancient Celts* (Penguin, 1999)

B. Cunliffe, *Facing The Ocean* (Oxford University Press, 2001)

K. Davies, *Sequence: Inside the Race for the Human Genome* (Weidenfeld and Nicolson, 2001)

J. Haywood (ed.), *The Penguin Historical Atlas of the Vikings* (Penguin, 1995)

D. Hey, *The Oxford Guide to Family History* (Oxford University Press, 1993)

D. Hey, *The Oxford Companion to Local and Family History* (Oxford University Press, 1996)

D. Hey, *Family Names and Family History* (Hambledon Press, 2000)

M. Jones, *The Molecule Hunt* (Penguin, 2002)

S. Jones, *In the Blood: God, Genes and Destiny* (Doubleday, 1997)

S. Jones, *The Language of Genes* (Flamingo, 2000)

S. Jones, *Y: The Descent of Men* (Little, Brown, 2002)

G. Lasker, *Surnames and Genetic Structure* (Cambridge University Press, 1985)

S. Mithen, *After the Ice* (Weidenfeld and Nicolson, 2003)

S. Olson, *Mapping Human History* (Bloomsbury, 2002)

S. Oppenheimer, *Out of Eden* (Constable and Robinson, 2003)

E. Passarge (ed.), *Color Atlas of Genetics* (Georg Thieme Verlag, 2001)

P. Reaney (ed.), *A Dictionary of English Surnames* (Oxford University Press, 1997)

G. Redmonds, *Surnames and Genealogy* (Federation of Family History Societies, 2002)

G. Redmonds, *Christian Names in Local and Family History* (The National Archives, 2004)

J. Richards, *Blood of the Vikings* (Hodder and Stoughton, 2001)

M. Ridley, *Genome: The Autobiography of a Species* (Fourth Estate, 1999)

C. Rogers, *The Surname Detective* (Manchester University Press, 1995)

A. Savin, *DNA for Family Historians* (Alan Savin, 2000) <http://savin.org>

B. Sykes, *The Seven Daughters of Eve* (Bantam, 2001)

B. Sykes, *Adam's Curse* (Bantam, 2003)

S. Wells, *The Journey of Man: A Genetic Odyssey* (Penguin, 2003)

A list of dozens of downloadable academic papers and other online resources is maintained at **<www.DNAandFamilyHistory.com>** including:

M. Richards *et al.* (2000), 'Tracing European founder lineages in the Near Eastern mtDNA pool', *American Journal of Human Genetics*, 67, 1251–76.

P. Underhill *et al.* (2000), 'Y chromosome sequence variation and the history of human populations', *Nature Genetics*, 26, 358–61.

The Y Chromosome Consortium (2002), 'A nomenclature system for the tree of human Y-chromosomal binary haplogroups', *Genome Research*, 12, 339–48.

INDEX

Numbers in *italics* indicate
Figures; those in **bold** indicate
Tables.

adenine (A) 10, *10*, 157
Adobe 146
adoption 76
African Ancestry 95
African Lineage Database 107
aliases 76-7
allele values 9, 12, 14, 50, 54, 59,
    157
alleles 106, 157, 162
Allred DNA project 73-4
Alpin, King 91
*American Journal of Human
    Genetics* 153
American Society of Human
    Genetics 37
Americas 27, *27*, 28, 29-31, 32,
    56
Ancestry by DNA 108
Ancestry.com 106, 117
Argyll, Duke of 72
Atlantic Modal Haplotype 130,
    157
Aurignacian period 32
autosomal DNA 13, 14, 15, 157
    tests 14-15, 18, 95, 108
autosomes 157

Babur surname 93
background family data 137-8
background surname data 138
base pairs 10, *10*, 50, 157
BBC vii, 38
Beringia 28, 30, 157, 160
black American men 95
Black Death 45
Bolling study 65
Boyd study 72
Brazil study 84
Brigham Young University 55
Butler, J. 158, 163
byname 157

calibration issues 151
Cambridge Reference Sequence
    14, 34, 157
Campbell study 72
Cann, Rebecca 162
castes 68, 157
    see also clan and caste studies
Caucasus 22, 32, 35
cells 9, *9*, 11, 14, 157-8
censuses 41, 75
Chitpavans 92-3
chromosomes 9, *9*, 13, 158, 162,
    163
clades 11, 22, 31-4, *33*, 36, 37, *54*,
    58, 158, 162, 163
    see also haplogroups
clan and caste studies 88-96
clans 66-7, 158, 164

see also clan and caste studies
clerical mutation 57, 114, 158
climatic change 21-2, 23, **24**, 25
Cline, Sebastian 75
coding regions 158
Cohanim 67
consent form 120, 121
Cooper study 83
cytoplasm 9
cytosine (C) 10, *10*, 157

Data Protection Act 158
data see under presenting and
    publishing your study
databases see under Y-chromo-
    some test
'deep ancestry' 18, 19, 34, 35,
    94-5, 96, 99, 158
Dixit, Dr Jay 92
DNA (deoxyribonucleic acid) 8,
    10, 11, 158
DNA Heritage, Weymouth,
    Dorset 53, 54, 104, 105, 112,
    113
DNA results, analyzing 122-33,
    132
DNA signatures 1, 16, 50-54, 56,
    57, 58, 61, 63, 158, 163
    Allred DNA project 73-4
    clusters 65
    of the common ancestor at
        the head of a family tree 71
    and large family trees 118
    and low-resolution tests
        99-100
    null matches 77
    rare 109
    Stiddem DNA study 74
    very old familial connection
        101
DNA surname projects:
    Allred 73-4
    Bolling 65
    Boyd 72
    Brazil 84
    Campbell 72
    Donald 65, 85, 92
        see also Macdonald family
        and MacDougall family
    Graves 65, 85
    Guinness 89
    Long/Lang 83
    MacGregor 91
    Mumma 66, 73, 74, 75, 129,
        146
    O'Donaghue 89-90
    O'Rourke 89
    O'Shea 90
    Payne 84
    Pennington 84
    Pomeroy vii, 65, 66, 72-3,
        82-3, 85-7, 129, **136**, 138,
        139, **140**, 141, **141**, 143, 147,
        155
    Rice 84
    Smith 82
    Stiddem, 64, 74

Sykes vii, 15-16, 37, 38-9, 45,
    46, 81
Turner 82
Walker 65, 85
Wells 65, 85, 153
DNA surname studies 102-3,
    104, 116-21
    see also DNA results, analyz-
        ing; presenting and publish-
        ing your study
DNA Test Checklist 98-103, *98*
DNA testing 1-4, 11-17, 115,
    149-50, 158, 163
    see also DNA results,
        analyzing
DNA testing companies 34, 150
DNA testing company selection
    104-15, *111*, *113*, **114**
DNAPrint Genomics 108
documentation 85-7, 119-20
    see also presenting and
        publishing your study
Donald clan 91
Donald study 65, 85, 92
double helix 10, *10*, 158-9
double-step mutation 125, 159
DYS (DNA/Y-chromosome/
    Single-copy sequence) 159

Edmund Rice (1638) Association
    84
electoral roles 41
email 117-18, 121, 153
emigrant/immigrant studies 83
Emperor of India Surname
    Project 93
ethnic group studies 93
'extra-paternity event' 76

Family Tree DNA, Houston,
    Texas 54, 57, 92, 104, 105-6,
    112, 113, 117, 119, 150
family trees, 17, 55, 64, 71, 72, 73,
    77, 86, 101, 123, 125-6, 145
future of genetic genealogy
    148-54

Gedcom 55, 137, 151
GenByGen, Göttingen 107
gene groups, classifying 31
genealogical magazines 119
Genealogy-DNA List Task Force
    117
genetics 160
    genes 9, 11, 13, 22, 159
    genetic bottleneck 21, 29, 30,
        45, 73, 159
    genetic clan 159
    genetic diversity 21, 159
    genetic drift 159
    genetic families 17, 64, 89,
        140, 159
    genetic refuges 29, 159-60
    genetic trees, and family trees
        77
GeneTree 107
GenForum 117

Genghis Khan 94, 107, 149
genome 160
Genome Research 164
Genuki vii
GeoGene 108
geographical mapping 144
Ghazi II, Emperor of India 93
Graves, Ken 85
Graves study 65, 85
Gravettian period 32
Greenshields, Alastair 105
Greenspan, Bennett 105
Gregor 91
guanine (G) 10, *10*, 157
Guild of One-Name Studies
(GOONS) 85, 116
Guinness family 89

Hammer, Dr Michael 31, 105
Hanson, John 90
haplogroups 11, 22, 33, 34, 54,
99, 160, 163
see also clades
haplotypes 12, 16-17, *143*, 160,
163
associated with the Jewish
priestly caste 67
clusters 145
distribution 36, 47, 123, 129
DNA signature 64
European 95
fit 123
founding 129
frequency 123
and illegitimacy 77
invader/earlier inhabitant
37
labels *143*
mapping 140, **141**
overall geographical
distribution 52
phylogenetic 142
publication of regional data
150-51, 154
rare 56-7, 75
reassessment 58, 61
Hay (previously Boyd), James,
15th Earl of Errol 72
Hey, Professor David 39, 40, 43,
44-5, 90
Hill study 65, 85
Homo erectus 30
Homo neanderthalensis 30
Homo sapiens 2, 22, 30, 161
Howard University, Washington
107
Huddersfield University 32
HUGO (Human Gene Nomen-
clature Committee) 160
human migration 18, 20-35, **24**,
45
Huns 93

Ice Ages 23, 25, 32
the last 29, 30, 35, 36
illegitimacy 71-5, 152
interglacial optima 22, 25, 160

International Forensic Y-User
Group 56
interpretation process 126-7
interstadials 23, 25, 160
Irish clans 66, 67, 88, 89-90, 96
ISO 17025-certified 106, 113

Jewish priestly caste 67, 94-5
junk DNA 9, 11, 13, 153, 160

Kayser, Manfred 153
Kittles, Dr Rick 107
Klein, Johann Heinrich 75
Kokanastha Brahmin 92
Kramwinckel, Peter 119

Lasker, Gabriel Ward 37
Last Glacial Maximum (LGM)
29, 31, 160
Little, Captain Daniel 75
locus, loci 12, 13, 50, 160, 162
Long/Lang project 83

MacAlistair family 91
McCartan clan 89
MacDonald family 91, 92,
129-30
MacDougall family 91
MacGregor, Clan 91
McTiernan study 89
Magennis family 89
maiden names 137
markers 13-14, 16, 31, 160
configuration of 149
multi-copy 14, 161
mutation 59, 60
number measured 108-9
paired 14, 162
refinement of marker sets
151
test resolution 123
marketing of DNA study 118-19
maternal line ancestry 13, 14, 21,
161
mtDNA 'deep ancestry' test
34
Max Planck Institute of
Evolutionary Genetics,
Leipzig 153
medieval warm period 25
Melungeons 68, 93
Microsoft Excel 147
Microsoft Word 146, 147
Middle Upper Palaeolithic period
32
migration *see* human migration
mitochondria 9
non-recombining 15
mitochondrial DNA (mtDNA)
13, 14, 32, 33, 95, 160-61
mitochondrial DNA (mtDNA)
tests 13, 14, 18, 21, 23, 95,
99, 108
modal values 14, 50, 54, 161
'Modal versus Multiple Origin'
debate 87, 155
Mongols 93

Most Recent Common Ancestor
(MRCA) calculation 60, 131,
132, 133, *155*, 161, 163
mtDNA *see* mitochondrial DNA
Mughul surname 93
multi-copy markers 14, 161
multi-surname studies 82-3
Multiregional Thesis 161, 162
Mumma, Doug 114, 146, 158
Mumma surname project 66, 73,
74, 75, 129, 146
mutation 11, 13, 76, 161, 163
clerical 57, 114, 158
dating of 21
double-step 125, 159
random 14
rates 60, 123, 127, 133, 150
single-step 59, 125, 143, *143*,
163
thinking about 124-5
types of 58-9

National Health Service (NHS)
Register 138
National Institute of Standards
and Technology *see* NIST
Native Americans 108
Nature 162
NIST (National Institute of
Standards and Technology)
161
non-coding region *see* junk DNA
'non-paternity event' 76, 77
Norman Conquest 39
Norman period 39, 43, 72
Norse ancestry 91, 92, 94
nuclear DNA 161
nucleotides 10, *10*, 161

O'Donaghue study 89-90
Oppenheimer, Dr Stephen 26-7,
28, 31, 40, 162
oral history 72-3, 86
O'Rourke study 89
O'Shea study 90
'Out of Africa' thesis 21, 22, 25,
28, 161, 162
Outlook Express 119
Oxford Ancestors 107, 112
Oxford University vii, 15, 32, 37,
107

PA Deutsch Ethnic Group DNA
project 93
paired markers 14, 162
palaeoanthropologists 2, 21
palaeographers 2
'part-takers' 91
Passarge, E. 158
paternal-line ancestry 21
paternity tests 2, 8
patronymic 162
Payne study 84
PDF software 146, 147
Pennington Family History
Association 84
Pennington study 84

Pennsylvanian Germans
(Pennsylvanian Dutch) 93
phylogenetics 31
phylogenetic charts 141, 142,
143, 144
phylogenetic tree 22, 162
phylogeography 12, 14, 15, 20,
31, 162
Picts 90
poll tax 39
polymorphism 11, 13, 14, 162,
163
Pomeroy DNA surname study
vii, 65, 66, 72-3, 82-3, 85-7,
129, **136**, 138, **139, 140**, 141,
141, 143, 147, 155
Pomeroy, Eltweed 73
Pomeroy, Henry 86-7
Pomery, and variants 38, 40
population history 152
presenting and publishing your
study 134-47, **139, 140, 141,**
142, 143
property inheritance 76
protein 162

questionnaires 118

race identification 2
Reaney, Percy Hide: *A
Dictionary of English
Surnames* 41-2
recalibration 57
recombination 14-15, 157, 162-3,
164
Redmonds, Dr George 42, 43, 44
regional and locality studies 93-4
Relative Genetics, Salt Lake City,
Utah 55, 104, 106-7, 112,
113
resolution 51-2, 81-2, 103, 128,
149, 154, 163
Rice Association, Edmund 84
Rice study 84
Richards, Dr Martin 32, 33
Roots for Real 108
Rootsweb vii, 90, 117
Freepages 119
Roper, David 75
Rose study 65, 85
rule-of-thumb assessments
131-3, *132*

salinity 21, 26
Scottish clans 66, 67, 88, 90-92,
94, 96
Scots 90
Scots-DNA list, Rootsweb 90
sea levels 20, 21, 25-6
septs 90, 91
sex chromosomes 163
Shetland Islands project 94
single nucleotide polymorphism
see SNP
single-ancestor studies 70-78
single-step mutation 59, 125,
143, *143*, 163

Smith project 82
SNPs (single nucleotide poly-
morphisms) 11, 12, 16, 22,
34, 149, 153, 160, 163
Somerled 91, 92, 129-30
Sorenson Genomics, USA 105,
106, 107, 113
Sorenson Group 55
Sorenson, James 106
Sorenson Molecular Genealogy
Database (SMGD) 53, 55-6
Sorenson Molecular Genealogy
Foundation 55, 56, 107, 112
Soundex 64, 79, 84
Stanford University, California
33
Steadham, Richard 64
Steedman, John 74
step relationships 76
Stiddem DNA study 64, 74
Stiddem, Timen 64, 74
Stidham, Samuel 64
Stidham family 74
STRs (short tandem repeats)
11-14, 16, 34, 50, 149, 151,
160, 162, 163-4
surname studies 3, 15-16, 79-87
surnames 36-47
breakdown of surname transmission process 153
categorization 123
change of 76-7
common 81-2, 102
fit 123
frequency 123
handed down from one
generation to the next 12
multiple-ancestor 38-9, 42-6,
69, 80, 152
naturalization 75, 83
rare 102
single-ancestor 38, 43-7, 69,
80, 152
spelling variations 39-40, 77
thinking about 124
Sykes, Professor Bryan vii, 15,
37, 81, 92, 107
Sykes surname study vii, 15-16,
37, 38-9, 45, 46, 81

tandem repeat *see* STR; VNTR
tax lists 41
test resolution 123, 124
Thomas, Dr Mark 37
thymine (T) 10, *10*, 157
timeline charts 144-5
Timen Stiddem Society 64
Timur surname 93
tool technology 32
Trinity College Dublin 89
Turner study 82

Underhill, Dr Peter 33
University College London
(UCL) 37
University of Arizona 31, 105,
106

University of California, Santa
Cruz 153
unmarried mothers 76
Upper Palaeolithic period 32

Vikings 18, 25, 36, 37, 91, 92, 94,
95
VNTR (variable number of
tandem repeats) 12, 164

Walker study 65, 85
Walsh, Dr Bruce 105
web pages 119
Wells, Orin 85, 153
Wells study 65, 85, 153
Welsh Patronymics project 94
West Country 130, 137
'wet science' 151

X-chromosomes 163

Y-Chromosome Consortium
(YCC) 31, 34, 105, 106, 149,
164
Y-Chromosome Haplotype
Reference Database
(YHRD; formerly known as
Y-STR database) 53, 56-7,
101, 125, 130, 151
Y-chromosome study scenarios
62-9
Y-chromosome test (Y-line,
Y-test) 12-14
comparing Y-chromosome
results 52
costs 100
'deep ancestry' 34, 35
a first glance at the results
16-17
a male-only test 1, 12, 13
new knowledge earned 58-9
rates of mutation 60
reading a result 50-51
resolutions 16, 19, 51-2
tracing paternal descent from
a single man 23
types of mutation 58-9
use by family historians 17
using the free-to-use online
databases 52-7, 53
a warning 57-8
Y-chromosomes 13, 15, 164
YBase database 53-4, 53, 57, 105
YSearch database 53, 54-5, 57,
106

CPSIA information can be obtained at www.ICGtesting.com
Printed in the USA
BVOW04s0240100614

355931BV00017B/329/P

9 781550 025361